Lecture Notes in Mathematics

Edited by A. Dold and B. Eckmann

490

The Geometry of Metric
and Linear Spaces

Michigan 1974

Springer-Verlag
Berlin · Heidelberg · New York

Lecture Notes in Mathematics

continuation on page 245

Lecture Notes in Mathematics

Edited by A. Dold and B. Eckmann

490

The Geometry of Metric and Linear Spaces

Proceedings of a Conference Held at Michigan State
University, East Lansing, June 17–19, 1974

Edited by L. M. Kelly

Springer-Verlag
Berlin · Heidelberg · New York 1975

Editor
Prof. L. M. Kelly
Department of Mathematics
Michigan State University
East Lansing
Michigan 48824/USA

Library of Congress Cataloging in Publication Data

Main entry under title:

The geometry of metric and linear spaces.

    (Lecture notes in mathematics ; 490)
    Bibliography: p.
    Includes index.
    1. Normed linear spaces--Congresses. 2. Inner
product spaces--Congresses. 3. Metric spaces--
Congresses. 4. Convex sets--Congresses.
5. Geometry--Congresses. I. Kelley, Leroy Milton,
1914-   II. Series: Lecture notes in mathematics (Berlin) ; 490.
QA3.L28 no. 490 [QA322.2]   510'.8s [515'.73]
                                 75-33966

AMS Subject Classifications (1971): 50-02, 50 A05, 50 A10, 50 C05,
52 A05, 52 A20, 52 A25, 52 A35, 52 A45, 46 B10, 46 C05.

ISBN 3-540-07417-1 Springer-Verlag Berlin · Heidelberg · New York
ISBN 0-387-07417-1 Springer-Verlag New York · Heidelberg · Berlin

PREFACE

One of the purposes for organizing this conference was to give formal recognition to the contributions of Professor Leonard Mascot Blumenthal to the foundational study of the geometry of metric spaces. The participants presented Professor Blumenthal with a scroll dedicating the conference to his pioneering work in the field. These proceedings are likewise dedicated to him.

The papers in this volume are essentially elaborations of those presented to the conference either orally or by title. In one or two instances they are notes inspired by such papers.

The papers presented seemed to fall naturally into three categories, (a) those concerned with purely metric questions, (b) those concerned with metric geometry in general linear spaces, and (c) those concerned with the geometry of Minkowski spaces or convexity. We have organized the papers here in that order.

Finally we present a list of problems suggested by the participants and edited by R.C. Guy.

Table of Contents

CONFERENCE ON THE GEOMETRY OF METRIC AND LINEAR SPACES

PARTICIPANTS

Professor J. Adney
Michigan State University
East Lansing, MI 48824

Professor J.R. Alexander, Jr.
University of Illinois
Urbana, IL 61801

Professor E.Z. Andalafte
University of Missouri at S.L.
St. Louis, MO 63121

Professor David Barnette
UCD-Davis
Davis, CA 95616

Professor L.M. Blumenthal
University of Missouri
Columbia, MO 65201

Professor W. Bonnice
University of New Hampshire
Durham, NH 03824

Dr. Marilyn Breen
University of Oklahoma
Norman, Oklahoma 73069

Professor G.D. Chakerian
U.C.D.
Davis, California 95616

Professor James Chew
Michigan State University
East Lansing, MI 48824

Professor M.M. Day
University of Illinois
Urbana, IL 61801

Professor M. Edelstein
Dalhousie University
Halifax, Nova Scotia
Canada

Professor J.S. Frame
Michigan State University
East Lansing, MI 48824

Professor R. Freese
St. Louis University
St. Louis, MO 63103

Professor Branko Grunbaum
University of Washington
Seattle, WA 98195

Professor Richard Guy
University of Calgary
Calgary (44) Canada

Professor William Hare
Clemson University
Clemson, SC 29631

Professor David Kay
University of Oklahoma
Normal, OK 73069

Professor John Kelly
Arizona State University
Tempe, AZ 85281

Professor L.M. Kelly
Michigan State University
East Lansing, MI 48824

Professor J. Kinney
Michigan State University
East Lansing, MI 48824

Professor W.A. Kirk
University of Iowa
Iowa City, Iowa 52240

Mr. Murray Klamkin
Ford Motor Science Lab.
POB 2053
Dearborn, MI 48121

Professor V. Klee
University of Washington
Seattle, WA 98195

Dr. J.B. Kruskal
Bell Telephone Labs
Murray Hill, NJ 07974

Professor G. Ludden
Michigan State University
East Lansing, MI 48824

Professor John Oman
Wisconsin State University
Oshkosh, WI 54901

Professor C.N. Petty
University of Missouri
Columbia, MO 65201

Professor J.R. Reay
Western Washington St.
College
Bellingham, WA 98225

Professor G.T. Sallee
UCD, Davis
Davis, CA 95616

Professor J.J. Seidel
Thomas J. Watson Research Center
P.O. Box 218
Yorktown Hts., NY 10598

Professor Andrew Sobczyk
Clemson University
Clemson, SC 29631

Professor S.V. Sreedharan
Michigan State University
East Lansing, MI 48824

Dean W.L. Stamey
College of Arts and Science
Eisenhower Hall
Kansas State University
Manhattan, KS 66502

Professor K.B. Stolarsky
University of Illinois
Urbana, IL 61801

Professor K. Sundaresan
University of Pittsburgh
Pittsburgh, PA 15213

Professor Joseph Valentine
Utah State University
Logan, UT 84321

Professor Dorothy Wolfe
245 Hathaway Lane
Wynnewood, PA 19096

Michigan State University - East Lansing, Michigan

Department of Mathematics

Conference: The Geometry of Metric and Linear Spaces

Program

Monday, June 17 - Room B-102 Wells Hall

A.M.	9:15	Professor L.M. Blumenthal, University of Missouri	- A review of some metric characteri- zations of euclidean space.
	10:15-10:45	Recess and Refreshments	
	10:45-11:30	Professor W.A. Kirk, University of Iowa	- Geometric fixed point theory and inwardness conditions.
	11:30-12:15	Professor J.J. Seidel, Technological University Eindhoven	- The state of affairs in elliptic space.
P.M.	2:15- 3:15	Professor John B. Kelly, Arizona State University	- Hypermetric spaces.
	3:15- 3:45	Recess and Refreshments	
	3:45- 4:30	Professor K.B. Stolarsky, University of Illinois	- Discrepancy and sums of distance between points of metric space.
	4:30- 5:00	Professor J. Valentine, Utah State University	- Angles in metric spaces.
P.M.	8:00- 9:00	SPECIAL LECTURE	- ROOM B-102 Wells Hall

Dr. J.B. Kruskal, Bell Telephone Labs

"Multidimensional scaling and clustering:
Embedding of noisy data in metric spaces."

	9:00-	Reception	V.G. Grove Mathematics Library D-101 Wells Hall

Tuesday, June 18 - Room B-102 Wells Hall

A.M.	9:15-10:15	Professor M.M. Day University of Illinois	- Mimicry in normed spaces.
	10:15-10:30	Recess and Refreshments	
	10:30-11:15	Professor M. Edelstein, Dalhousie University	- Fixed point theory in linear spaces.

11:15–12:00	Professor J. Oman, Wisconsin State University	– Characterization of Hilbert space.
P.M. 2:00–3:00	Professor V. Klee, University of Washington	– Convex polytopes and mathematical programming.
3:00–3:15	Recess and Refreshments	
3:15–4:00	Professor David Barnette, University of California Davis	– Generalized Combinatorial Spheres and Facet Splitting.
P.M. 7:00	DINNER – University Club	

Wednesday, June 19 – Room A-304 Wells Hall

A.M. 9:15–10:15	Professor B. Grunbaum, University of Washington	– Polygons
10:15–10:45	Recess and Refreshments	
10:45–11:30	Professor C.N. Petty, University of Missouri	– Intersectional configurations of convex sets.
11:30–12:15	Professor Dorothy Wolfe, Widener College	– Metric dependence: Some geometric consequences.
P.M. 2:00–2:45	Professor G.D. Chakerian, U.C.D. – Davis	– Covering Spaces with Convex Bodies.
2:45–3:30	Professor K. Sundaresan, University of Pittsburgh	– Smoothness classification of reflexive spaces.
3:30–4:00	Professor G.T. Sallee	– A Helly Type Theorem for widths.

FOUR-POINT PROPERTIES AND NORM POSTULATES

Leonard M. Blumenthal
University of Missouri

1. Introduction

This paper exhibits and exploits for the frist time the relations between two problems that arose about forty years ago, attracted numerous investigators, and continue to be studied, mostly without reference to their inter-connections.

Those problems are (1) to determine conditions on the six mutual distances of four points of a complete, metrically convex and externally convex metric space (denoted here by $\mathcal{M}$) that are necessary and sufficient in order that the metric of $\mathcal{M}$ may be euclidean, and (2) to ascertain conditions on the norm of a real Banach space (denoted here by $\mathcal{B}$) that are necessary and sufficient to permit defining an inner product in $\mathcal{B}$, which is connected with the norm in the usual manner. We refer to the desired conditions on the six mutual distances of four points of $\mathcal{M}$ as four-point properties, and call the sought-for conditions on the norm of $\mathcal{B}$, norm postulates. One four-point property was established by W.A. Wilson in 1932, and a norm postulate follows from Fréchet's paper of 1935, and by the well-known result of Jordan and von Neumann of the same date (explicit references will be provided later).

The connection between the two problems stated above is made clear by noting that (a) every Banach space $\mathcal{B}$ becomes a complete, metrically convex and externally convex metric space $\mathcal{M}$ upon defining the distance xy of any two elements (points) x, y by $xy = \|x - y\|$, (b) norm postulates on $\mathcal{B}$ may be interpreted as conditions on the mutual distances of the points of certain finite subsets of $\mathcal{B}$ (quadruples of points in those cases we shall consider here), while four-point properties applied to Banach spaces yield norm postulates, and (c) the existence f an inner product in β is necessary and sufficient for the metric of β to be euclidean.

It follows that each four-point property which establishes the euclidean nature of the metric of a complete, convex, externally convex metric space may be applied to a Banach space to yield, <u>as a corollary</u>, a norm postulate which is a necessary and sufficient condition for defining an inner product in the space; while, on the other hand, each norm postulate that permits an inner product to be defined in a Banach space <u>suggests</u> a four-point condition that <u>might</u> imply the euclidean nature of the metric of a complete, convex, externally convex metric space. In case it does, the inner product theorem becomes an immediate corollary of the metric space theorem.

In this paper it is shown that (i) Fréchet's inner product theorem of 1935 is a corollary of Wilson's four-point theorem of 1932; (ii) Kasahara's theorem of 1954 (improved by Day in 1959) follows directly from the euclidean <u>weak</u> four-point property introduced by the writer in 1935; (iii) the Jordan and von Neumann norm condition of 1935 suggested to the writer the euclidean <u>feeble</u> four-point property, which yields a characterization of euclidean spaces among complete, convex, externally convex metric spaces, and hence generalizes the Jordan - von Neumann theorem. Other results of this nature due to Day, Freese, Valentine and Andalafte are discussed, and further work along these lines is suggested by two conjectures.

The paper ends with a discussion of other approaches to such problems, in particular, with work by the writer on quasi inner product spaces.

2. Wilson's Euclidean Four-Point Property and Fréchet's Norm Postulate

In 1932 W.A. Wilson [17] achieved a breakthrough in solving the problem of characterizing metrically n-dimensional euclidean space $\mathcal{E}_n$ among the class of all complete, convex, externally convex metric spaces by showing that the assumptions of the congruent imbeddability in $\mathcal{E}_k$ of each $(k + 1)$-tuple of points of the space, $(3 \leq k \leq n)$, which featured the earlier solution of the problem obtained by Menger, could be replaced by assuming merely that each four points of the space be congruently imbeddable in $\mathcal{E}_3$, no matter what the dimension of the euclidean space may be. Now, a metric quadruple p_1, p_2, p_3, p_4 is congruent with a quadruple of $\mathcal{E}_3$ if and only if the determinant $D(p_1, p_2, p_3, p_4)$, obtained from the determinant $|p_i p_j^2|$ $(i, j = 1, 2, 3, 4)$, by bordering with a row and a column of 1's and intersecting element 0, is non-negative. We shall refer to the inequality $D(p_1, p_2, p_3, p_4) \geq 0$, or to its geometrical equivalent, as the euclidean four-point property.

In 1935 Fréchet [9] proved that an inner product may be defined in a Banach space if and only if, for each three points x_1, x_2, x_3 of $\mathcal{B}$, the quadratic form

$$Q = (1/2) \sum_{i,j=1}^{3} [\, \|x_i\|^2 + \|x_j\|^2 - \|x_i - x_j\|^2] u_i u_j$$

be non-negative for all real u_1, u_2, u_3. It follows that a norm postulate for $\mathcal{B}$ to be an inner product space is

$$\det[\, \|x_i\|^2 + \|x_j\|^2 - \|x_i - x_j\|^2 \geq 0 \quad (i, j = 1, 2, 3).$$

Examining the geometrical significance of his norm postulate, Fréchet showed that it is equivalent to the congruent imbedding in $\mathcal{E}_3$ of every four points of $\mathcal{B}$; that is, that $\mathcal{B}$ is an inner product space if and only if it has the euclidean four-point property. But then his result follows at once from the more general theorem proved by Wilson three years earlier.

The connection between Fréchet's quadratic form or norm criteria for the imbedding of four points of β in δ_3 and the determinant criteria stated above is made clear by the relation

$$D(\theta, x_1, x_2, x_3) = 8|x_{ij}| = 8 \det Q,$$

where $x_{ij} = (1/2)[\|x_i\|^2 + \|x_j\|^2 - \|x_i - x_j\|^2]$, $(i, j = 1, 2, 3)$.

In later sections much simpler norm postulates will appear as the results of establishing stronger theorems of Wilson's type by restricting the class of quadruples assumed imbeddable in euclidean space.

3. <u>The Weak Euclidean Four-Point Property and the Kasahara-Day Norm Postulate</u>

It was shown by the writer (1935, 1953) [2,4] in proofs which, unlike Wilson's made no use of Menger's imbedding theorems, that the metric of a complete, convex, externally convex metric space is euclidean if and only if every quadruple of its points <u>containing a linear triple</u> is congruently contained in the plane δ_2. That the class of imbeddable quadruples could be so restricted turned out to be quite useful. Let us apply this result to Banach spaces.

It is easily seen that any metric space with the weak euclidean four-point property has unique metric lines, and consequently, if three points of β (with that property) are linear they may be denoted $x, y, \lambda x + (1 - \lambda)y$ with $0 < \lambda < 1$. Then $\theta, x, y, \lambda x + (1 - \lambda)y$ are congruently imbeddable in δ_2 and so $D(\theta, x, y, \lambda x + (1 - \lambda)y) = 0$.

Developing the determinant yields the norm postulate

$$\|\lambda x + (1 - \lambda)y\|^2 + \lambda(1 - \lambda)\|x - y\|^2 = \lambda\|x\|^2 + (1 - \lambda)\|y\|^2,$$
$$0 < \lambda < 1,$$

which is the simplification effected by Day [8] in 1959 of a condition established by Kasahara in 1954 [12].

4. The Feeble Euclidean Four-Point Property and the Jordan - von Neumann Norm Postulate

Having successfully moderated the euclidean four-point property demand to weak euclidean, it was natural for the writer to seek additional restrictions on the class of quadruples assumed imbeddable in $\mathcal{E}_2$. A suggestion of a possible direction to investigate was provided by the result obtained by Jordan and von Neumann in 1935 [11] which has since become the best known of all norm postulates that permit a Banach space to admit an inner product; namely, for all $x, y \in \mathcal{B}$,

$$\|x + y\|^2 + \|x - y\|^2 = 2(\|x\|^2 + \|y\|^2).$$

It is convenient for our purposes to write that norm postulate

$$(*) \quad \|(x + y)/2\| = (1/2)[2\|x\|^2 + 2\|y\|^2 - \|x - y\|^2]^{1/2},$$

in which form it is seen to express the length of a median of a triangle as the same function of its sides that it is in the euclidean plane. Hence $(*)$ implies that the quadruple $\theta, x, y, (x + y)/2$ is congruently imbeddable in $\mathcal{E}_2$.

Before passing to the metrization of the Jordan - von Neumann norm postulate $(*)$, which is our first objective, it is important to observe that $(*)$ is equivalent to

$(\#)$ if t denotes any metric middlepoint of x, y, then
$$\|2t\|^2 + \|x - y\|^2 = 2\|x\|^2 + 2\|y\|^2.$$

For since $(x + y)/2$ is a metric middlepoint of x, y, clearly $(\#)$ implies $(*)$. On the other hand, let t denote any metric middlepoint of x, y; that is, $\|x - t\| = \|y - t\| = (1/2)\|x - y\|$.

Applying (*) to the points x - t, y - t yields

$$\|x - t + y - t\|^2 + \|x - y\|^2 = 2\|x - t\|^2 + 2\|y - t\|^2$$
$$= \|x - y\|^2 .$$

Hence $\|x + y - 2t\| = 0$, and so $2t = x + y$. Substitution in (*) gives (#).

It is now clear that the metrization of the Jordan - von Neumann norm postulate may be stated in the following form.

Each metric quadruple p,q,r,s with q a middlepoint of p,r is congruently imbeddable in $\mathcal{E}_2$.

Since this four-point property is a priori weaker than the weak euclidean four-point property, we refer to it as the feeble euclidean four-point property.

The writer showed in 1955 [5] that any complete, convex, externally convex metric space with the feeble euclidean four-point property has a euclidean metric. From that result the Jordan - von Neumann norm postulate theorem follows as an abvious corollary. In this case, however, unlike the two preceding ones, the norm postulate inner product result was established before the corresponding four-point property theorem and, indeed, furnished its motivation.

5. Day's Queasy Four-Point Property

Though not quite in the program of specializing the structure of the class of quadruples assumed to be imbeddable in $\mathcal{E}_2$, Day established in 1959 [8] an a priori weakening of the feeble euclidean four-point property which he called the queasy euclidean four-point property. It assumes that p,r ϵ $\mathcal{M}$, (p $\neq$ r), implies the existence of q ϵ $\mathcal{M}$ such that (1) pq + qr = pr, (p $\neq$ q $\neq$ r) and (2) for every s ϵ $\mathcal{M}$, p,q,r,s are congruently imbeddable in $\mathcal{E}_2$. He showed if $\mathcal{M}$ has the queasy property then $\mathcal{M}$ has the feeble euclidean four-point property, and consequently, by the theorem of Section 4, $\mathcal{M}$ has euclidean metric.

Applying this result of Banach spaces gives a still better form of Kasahara's norm postulate (Section 2):

For each pair x,y of points of $\mathcal{B}$ there is a number λ, depending on x,y such that $0 < \lambda < 1$ and

$$\|\lambda x + (1 - \lambda)y\|^2 + \lambda(1 - \lambda)\|x - y\|^2 + \lambda\|x\|^2 + (1 - \lambda)\|y\|^2.$$

6. Freese's External Isosceles Feeble Four-Point Property

Continuing the program of putting restrictions on the quadruples assumed imbeddable in $\mathcal{E}_2$, Freese showed in 1968 [10] that the euclidean nature of the metric of a complete, convex, externally convex metric space followed by assuming the congruent imbeddability in the euclidean plane of all quadruples p,q,r,s of the space such that $pq + qr = pr$, $pq = 2qr$, and $ps = qs$. His proof consists in showing that such a space has the weak euclidean four-point property (Section 3).

Applying his result to Banach spaces, he obtained what appears to be a new norm postulate for an inner product; namely, $x,y \in \mathcal{B}$, $\|x - y\| = \|3x - y\|$ implies $\|y\|^2 = \|x - y\|^2 + 3\|x\|^2$.

7. The Isosceles Feeble Four-Point Property

A property that is more closely related to the feeble four-point property (Section 4) than the one considered in Section 6 is obtained by restricting the quadruples assumed imbeddable in Section 4 by requiring that the point s be equidistant from p and r. We say that $\mathcal{M}$ has the isosceles feeble euclidean four-point property provided $p,q,r,s \in \mathcal{M}$, $pq = qr = (1/2)pr$, $ps = rs$ implies p,q,r,s congruently imbeddable in the euclidean plane. Does this property suffice to make the metric of $\mathcal{M}$ euclidean?

An affirmative answer to this question seemed to be provided in 1971 [15] by an article of J.E. Valentine, but close examination of the paper shows that the proof is incomplete. Though a negative answer appears most unlikely, we must regard the question as an open one. A positive resolution of the matter would yield a broad extension of Day's norm postulate for an inner product in a Banach space, established in 1947; namely, $x, y \in \mathcal{B}$, $\|x\| = \|y\| = 1$ implies $\|x + y\|^2 + \|x - y\|^2 = 4$.

8.   Extension of the Aronszajn-Lorch Norm Postulate Criterion

In Section 4 it was remarked that the Jordan - von Neumann norm postulate implied that the length of a median of a triangle was assumed to be the same function of the sides of the triangle that it is in euclidean space. In 1935 [1] Aronszajn showed that an inner product results merely from supposing that the length of a median is a function of the lengths of the sides; that is, $x, y \in \mathcal{B}$ implies

$$\|(x + y)/2\| = (1/2)\emptyset(\|x\|, \|y\|, \|x - y\|) \geq 0,$$

with $\emptyset(a, 0, a) = a$.

This was re-discovered by Lorch in 1948 [13].

An interesting extension of Aronszajn's result to spaces $\mathcal{M}$ was obtained in 1973 by Valentine and Andalafte [16] as an easy application of a theorem by Busemann [7]. Those writers define a metric space to have the intrinsic feeble four-point property provided any congruence between two point triples p, q, r and p', q', r' of the space is extendible to a congruence between p, q, r, m and p', q', r', m', where m, m' are middlepoints of q and r, q' and r', respectively.

In addition to the properties of $\mathcal{M}$ assumed in Section 1, suppose $\mathcal{M}$ is finitely compact, has the two-triple property (that is, if any two of the point-triples contained in four points of $\mathcal{M}$ be linear, then the other two point-triples are linear also), and the intrinsic feeble four-point property. If $p, p' \in \mathcal{M}$, $(p \neq p')$,

put $E(p,p') = [x \in \mathfrak{M} | px = xp']$, and let q,r be distinct points
of $E(p,p')$. Then clearly $p,q,r \approx p',q,r$ and, by hypothesis,
this congruence can be extended to

$$p,r,r,m \approx p',q,r,m,$$

where m is the unique middlepoint of q and r. Hence $pm = mp'$,
and so $m \in E(p,p')$. Since $E(p,p')$ is closed, it follows easily
that $E(p,p')$ contains the (unique) segment joining q and r.
Then Busemann's theorem implies that $\mathfrak{M}$ is either euclidean or
hyperbolic.

To apply this result to the real Banach space $\mathcal{B}$, one observes that the intrinsic feeble four-point property (which is
equivalent to the Aronszajn-Lorch postulate) implies that metric
lines of $\mathcal{B}$ are unique (that is, $\mathcal{B}$ has the two-triple property)
and since $\mathcal{B}$ is not hyperbolic, it must be euclidean. Hence the
Aronszajn-Lorch norm postulate implies that $\mathcal{B}$ is an inner product space.

9. Two Open Questions

We have been concerned in the foregoing sections with norm
postulates that suffice to make any real Banach space an inner
product space and whose metrizations make any complete, convex,
externally convex metric space $\mathfrak{M}$ euclidean (of finite or infinite
dimension). Not all norm postulates are of that nature; for
example, the norm postulate

$$\|x\| \cdot \|y - z\| + \|z - x\| \geq \|z\| \cdot \|x - y\|, \quad x,y,z \in \mathcal{B},$$

which says that $\mathcal{B}$ is ptolemaic, implies that $\mathcal{B}$ is an inner
product space (Schoenberg [14]), but that ptolemaic inequality imposed on $\mathfrak{M}$ does not suffice to make $\mathfrak{M}$ euclidean.

Now it has been shown by several writers [6] that $\mathcal{B}$ admits an inner product provided $x, y \in \mathcal{B}$, $\|x + y\| = \|x - y\|$ implies $\|x + y\|^2 = \|x\|^2 + \|y\|^2$.

The metrization of this norm postulate is $p, q, r, s \in \mathcal{M}$, $pq = qr = (1/2)pr = qs$ implies p, q, r, s congruently imbeddable in the euclidean plane. We refer to this property as the <u>right triangle feeble euclidean four-point property</u>. It is, of course, a further restriction of the class of quadruples of $\mathcal{M}$ assumed imbeddable in the plane. <u>It is an open question whether that property suffices to make</u> $\mathcal{M}$ <u>euclidean</u>.

How far can one reasonably expect to proceed in specializing the structure of those quadruples assumed imbeddable in the euclidean plane with the result that the space $\mathcal{M}$ has the <u>equilateral feeble four-point property</u>; that is, assume that any four points p, q, r, s of $\mathcal{M}$ consisting of an equilateral triple $pq = qr = pr$ and s a middlepoint of q and r ($qs = rs = (1/2)qr$) are congruently imbeddable in the euclidean plane. It is an open question whether endowing $\mathcal{M}$ with that property implies that $\mathcal{M}$ is euclidean.

It has been suggested by Day and others that perhaps a counterexample might be the Minkowski plane of ordered pairs (x_1, x_2) of real numbers with unit circle a regular dodecagon. So far as the writer knows, the necessary computations have not yet been carried out.

Applied to a general Banach space, the equilateral feeble four-point property becomes

(*) $x, y \in \mathcal{B}$, $\|x\| = \|y\| = \|x - y\| = 1$ implies $\|x + y\| = \sqrt{3}$

The question then is whether or not (*) is a norm postulate that implies the existence in $\mathcal{B}$ of an inner product. An affirmative answer would be surprising.

10. Inner Product Banach Spaces in Terms of a Single Primitive Notion

In the preceding sections we have been concerned with various conditions imposed on the norm that permit defining an inner product in Banach spaces. This section deals with the reverse problem of imposing conditions on a rudimentary "inner product" defined in an abstract set Σ that permit defining "sum", "scalar multiplication" and "norm" in Σ so that Σ becomes a normed, linear space with an inner product, connected with the norm in the usual way. This problem was solved by the writer in 1950 [3] in the following manner.

Postulates for a Rudimentary Inner Product

To each pair of elements x, y of an abstract set Σ there is attached a real number (x, y) called a rudimentary inner product, in conformity with the following agreements:

Q_1 (Symmetry). If $x, y \in \Sigma$, then $(x, y) = (y, x)$.

Q_2 (Definiteness). For each element x of Σ, $(x, x) \geq 0$.

Q_3 (Identification). If $x, y \in \Sigma$ and $(x, x) = (x, y) = (y, y)$, then $x = y$.

It is pointed out that Q_3 is not a definition of equality of elements of Σ which, as a set, is already supplied with a criterion for equality of its elements. As its listing indicates, Q_3 is an assumption connecting the set-equality of elements x, y with the three numbers $(x, x), (x, y), (y, y)$ attached to the two elements as rudimentary inner products. On the other hand, Q_3 does, in the presence of Q_1, Q_2, S_2, E_1 actually define an equivalence relation in Σ.

The set Σ may then be referred to as a rudimentary inner product space.

Schwarz Postulates

If $x_1, x_2, \ldots, x_n \in \Sigma$, denote by $G(x_1, x_2, \ldots, x_n)$ the Gram determinant $|(x_i, x_j)|$, $(i, j = 1, 2, \ldots, n)$.

S_1. If $x_1, x_2 \in \Sigma$, then $G(x_1, x_2) \geqq 0$.

S_2. If $x_1, x_2, x_3 \in \Sigma$ and $G(x_1, x_2) = 0$, then

$G(x_1, x_2, x_3) \geqq 0$.

S_3. If $x_1, x_2, x_3, x_4 \in \Sigma$ and $G(x_1, x_2, x_3) = 0$, then

$G(x_1, x_2, x_3, x_4) \geqq 0$.

It is observed that S_2 and S_3 are conditional Schwarz inequalities, while S_1 applies to every pair of elements of Σ. The existence postulates that follow merely insure that Σ has enough elements for our purpose.

Existence Postulates

If $x_1, x_2, \ldots, x_n \in \Sigma$, denote by $B(x_1, x_2, \ldots, x_n)$ the symmetric determinant obtained by bordering $G(x_1, x_2, \ldots, x_n)$ with a row and column of 1's with intersection element 0.

E_1. There exists at least one element θ of Σ such that

$(\theta, x) = 0$ for each element x of Σ.

E_2. For each element x of Σ and each real number λ

there exists at least one element y of Σ such that

$G(x, y) = 0$ and $(x, y) = \lambda \cdot (x, x)$.

E_3. If $x, z \in \Sigma$ and $G(x, z) \neq 0$, there exists at least one

element y of Σ such that $B(x, y, z) = G(x, y, z) = 0$

and $G(x, y) = G(y, z)$.

It is proved that the space

$$\{\Sigma : Q_1, Q_2, Q_3, S_1, S_2, S_3, E_1, E_2, E_3\}$$

is a normed linear space, in which the rudimentary inner product is an ordinary inner product. The reader is referred to [3] for the definitions of sum, scalar multiplication, and norm, and for the details of the proof.

BIBLIOGRAPHY

[1]   N. Aronszajn, Caractérisation métrique de l'espace de Hilbert, des espaces vectoriels et de certains groupes métrique, Comp. R. Acad. Sci. Paris 201 (1935), 811-813; 873-875.

[2]   L.M. Blumenthal, Concerning spherical spaces, Amer. J. Math. 57 (1935), 51-61.

[3]   L.M. Blumenthal, Generalized euclidean space in terms of a quasi inner product, Amer. J. Math. 62 (1950), 686-698.

[4]   L.M. Blumenthal, Theory and applications of distance geometry, Clarendon Press, Oxford, 1953.

[5]   L.M. Blumenthal, An extension of a theorem of Jordan and von Neumann, Pac. J. Math. 5 (1955), 161-167.

[6]   L.M. Blumenthal, Note on normed linear spaces, Rev. R. Acad. Ciencias, Madrid 62 (1968), 307-310.

[7]   H. Busemann, On Leibnitz's definition of planes, Amer. J. Math. 63 (1941), 101-111.

[8]   M.M. Day, On criteria of Kasahara and Blumenthal for inner product spaces, Proc. Amer. Math. Soc. 10 (1959), 92-100.

[9]   M. Fréchet, Sur la définition axiomatique d'une classe d'espaces vectoriels distanciés applicables vectoriellement sur l'espace de Hilbert, Ann. Math. 36 (1935), 705-718.

[10] R.W. Freese, Criteria for inner product spaces, Proc. Amer. Math. Soc. 19 (1968), 953-958.

[11] P. Jordan and J. von Neumann, On inner products in linear metric spaces, Ann. Math. 36 (1935), 719-723.

[12] S. Kasahara, A characterization of Hilbert spaces, Proc. Jap. Acad. 30 (1954), 846-848.

[13] E.R. Lorch, On some implications which characterize Hilbert space, Ann. Math. 49 (1948), 523-532.

[14] I.J. Schoenberg, A remark on M.M. Day's characterization of inner product spaces and a conjecture of L.M. Blumenthal, Proc. Amer. Math. Soc. 3 (1952), 961-964.

[15] J.E. Valentine, On criteria of Blumenthal for inner product spaces, Fund. Math. 72 (1971), 265-269.

[16] J.E. Valentine and E.Z. Andalafte, Intrinsic four-point properties which characterize hyperbolic and euclidean spaces, Bull. Acad. Polonaise Sci. 21 (1973), 1103-1106.

[17] W.A. Wilson, A relation between metric and euclidean spaces, Amer. J. Math. 54 (1932), 505-517.

ON THE EQUILATERAL FEEBLE FOUR-POINT PROPERTY

Leroy M. Kelly
Michigan State University

In his paper Professor Blumenthal mentions an unchecked conjecture of Professor Day to the effect that a minkowski plane with
a regular dodecagon as unit circle satisfies the norm identity

$$(1) \quad \|x\| = \|y\| = \|x - y\| = 1 \Rightarrow \|x + y\| = \sqrt{3}.$$

More geometrically stated this says that the medians of equilateral
triangles of side length a are of length $\frac{\sqrt{3}}{2} a$ as they are in
E^2. Midpoint in this context is interpreted vectorially rather
than metrically.

Day proved that the norm identity

$$(2) \quad \|x\| = \|y\| = 1 \Rightarrow \|x - y\|^2 + \|x + y\|^2 = 4$$

in a linear space M is enough to insure that M is an inner product space. The Day conjecture, in effect, says that replacing (2)
by (1) is carrying things too far.

The proof of the validity of this conjecture is surprisingly
simple.

Let D be a regular dodecagon in E_2 with center O and
vertices A_i, $i = 1, 2, \ldots, 12$. If P and Q are two points in
the plane let $e(P, Q)$ denote the euclidean distance between them
and $m(P, Q)$ the distance in the minkowski metric in which D is
the unit circle. Let $P(\theta)$ denote a point on D with polar
coordinates $[e(O, P(\theta)), \theta]$.

Our problem now amounts to showing that if $m(P(\alpha), P(\beta)) = 1$
and K is the midpoint of $P(\alpha)$, $P(\beta)$ then $m(O, K) = \sqrt{3}/2$.

First note that $e(O, P(\theta)) = e(O, P(\theta + n \frac{\pi}{6}))$, $n = 1, 2, 3, \cdots$.
Now if $m(P(\alpha), P(\beta)) = 1$ the labeling can be so chosen that

$\alpha = \angle A_1 \ 0 \ P(\alpha) = \theta < \frac{\pi}{6}$ and $\beta < \frac{\pi}{6}$. It is clear in fact that subject to this restriction β is unique. We claim $\beta = \theta + \frac{\pi}{3}$.

$$e(0,P(\theta)) = e(0,P(\theta + \tfrac{\pi}{3})) \quad \text{so} \quad e(P(\theta),P(\theta + \tfrac{\pi}{3})) = e(0,P(\theta)).$$

$$m(P(\theta),P(\theta + \tfrac{\pi}{3})) = e(P(\theta),P(\theta + \tfrac{\pi}{3}))/e(0,P(\theta + \tfrac{2\pi}{3}))$$
$$= e(0,P(\theta))/e(0,P(\theta)) = 1.$$

Finally, $m(0,K) = e(0,K)/e(0,P(\theta + \tfrac{\pi}{6}))$
$$= \frac{\sqrt{3}}{2} \ e(0,P(\theta))/e(0,P(\theta)) = \frac{\sqrt{3}}{2} \quad \text{Q.E.D.}$$

Professor Blumenthal expressed interest in this conjecture because of its affinity to a property which he calls the equilateral feeble four-point property. A metric space is said to have the equilateral feeble four-point property if each of its subsets p,q,r,s with $pq = qr = pr$, $qs = rs = \frac{1}{2} qr$ is isometric to four points in E^2.

Blumenthal states that it is an open question whether a metric space with the equilateral feeble four-point property is isometrically embeddable in euclidean space.

The above example is not a counter-example to this conjecture since 0, $P(\frac{\pi}{12})$, $P(\frac{\pi}{4})$, $P(\frac{5\pi}{12})$ are such that

$$m(0,P(\tfrac{\pi}{6})) = m(0,P(\tfrac{5\pi}{4})) = m(P(\tfrac{\pi}{6}),P(\tfrac{5\pi}{4})) = 1$$

$$m(P(\tfrac{5\pi}{12}),P(\tfrac{\pi}{4})) = m(P(\tfrac{\pi}{4}),P(\tfrac{\pi}{12})) = \tfrac{1}{2} \ m(P(\tfrac{5\pi}{12}),P(\tfrac{\pi}{12})) = \tfrac{1}{2}$$

while $m(0,P(\frac{\pi}{4})) = 1$. That is, the minkowski plane with unit circle D does not have the equilateral feeble four-point property.

However it is clear that the same argument that works for the regular dodecagon will work equally well if the sides of the dodecagon are slightly bowed outward producing a convex regular "curvilinear" dodecagon. The crucial point in the argument is that $e(0,P(\theta)) = e(0,P(\theta + \frac{\pi}{6}))$.

The resulting minkowski planes do then provide examples of metric spaces with the equilateral feeble four-point property which are not euclidean.

HYPERMETRIC SPACES

John B. Kelly
Arizona State University

ABSTRACT

A metric space (M, ρ), is hypermetric if $\sum\limits_{1 \le i < j \le n} \rho(P_i, P_j) x_i x_j$ ≤ 0 for all positive n, all choices of $\{P_i\} \subseteq M$, $1 \le i \le n$ and all sets of integers $\{x_i\}$ such that $\sum\limits_{i=1}^{n} x_i = 1$. This paper describes some earlier work of the author on hypermetric spaces, some recent work of H. Witsenhausen and some simple results on hypermetric graphs. Hypermetric spaces had their origin in the study of the space Ω, of finite subsets of a set X, metrized by the cardinality of the symmetric difference. Ω is hypermetric and a general theorem shows that many of the more familiar metric spaces resemble Ω sufficiently to be hypermetric also. Thus S_n, the n-sphere with great circle distance, E_n, euclidean space and all one or two dimensional normed linear spaces are hypermetric. Distributive lattices are characterized among normed lattices by the property that the metric induced by the norm yields a hypermetric space.

A weaker condition, called quasi-hypermetricity is introduced. Witsenhausen showed that in normed linear spaces quasi-hypermetricity implies hypermetricity. This result has points of contact with the zonoid problem.

HYPERMETRIC SPACES

John B. Kelly
Arizona State University

Introduction

The resolution of many significant combinatorial problems depends upon our being able to decide whether there exists a family of finite sets with prescribed intersection cardinalities. For example, the still unsolved existence problem for finite projective planes is of this character. Any problem concerning intersection cardinalities can be replaced by an equivalent one concerning symmetric difference cardinalities. Inasmuch as the symmetric differences yield a metric on the underlying family of finite sets, one can then think about the combinatorial problem geometrically. It turns out that the induced metric has a property called "hypermetricity" which is shared by many other metric spaces of interest in geometry and analysis. In this article we shall attempt to summarize the work that has been done on hypermetric spaces.

Let k be a positive integer. A semi-metric space (M, ρ) will be said to be k-hypermetric if, given any set $\{P_1, \ldots, P_k, Q_1, \ldots, Q_{k+1}\}$ of $2k + 1$ points of M, one has

$$(1) \quad \sum_{1 \le i < j \le k} \rho(P_i, P_j) + \sum_{1 \le i < j \le k+1} \rho(Q_i, Q_j) \le \sum_{i=1}^{k} \sum_{j=1}^{k+1} \rho(P_i, Q_j).$$

Note that a 1-hypermetric space is simply a metric space. The k-hypermetric inequality gets stronger as k increases. One can show (4) that, if $k < \ell$, then every ℓ-hypermetric space is k-hypermetric and there exist k-hypermetric spaces which are not ℓ-hypermetric. A space that is k-hypermetric for all positive integers k is called hypermetric.

It is easy to see that (M, ρ) is hypermetric if and only if the following alternative condition is satisfied:

For all positive integers n, all choices of the set $\{P_1, P_2, \ldots, P_n\} \subseteq M$, and all sets of n <u>integers</u> $\{x_1, x_2, x_3, \ldots, x_n\}$ with

$$(2) \quad \sum_{i=1}^{n} x_i = 1$$

one has

$$(3) \quad \sum_{1 \leq i < j \leq n} \rho(P_i, P_j) x_i x_j \leq 0$$

(To show that (3) implies (1), set $n = 2k + 1$ and put $P_{k+1} = Q_1, \ldots, P_n = Q_{k+1}, x_1 + x_2 = \ldots = x_k = -1, x_{k+1} = \ldots = x_n = 1$. On the other hand, if we allow repetition of points in (1), we can readily see that the truth of (1) for all k yields (3).)

The following basic result in the theory of hypermetric spaces establishes their combinatorial significance.

<u>Theorem I</u>: Let (M, ρ) be a semi-metric space. Let (W, m) be a measure space and suppose that there is a mapping τ from M into the family of measurable subsets of W such that

$$\rho(P, Q) = m(\tau(P) \triangle \tau(Q))$$

for all (P, Q) in $M \times M$. (Here $\triangle$ denotes symmetric difference.) Then (M, ρ) is hypermetric.

Theorem I is proved in (4). In order to give the reader a taste of the kind of combinatorial reasoning used in the theory, we repeat the rather brief argument here.

<u>Proof</u>: Let $\{P_1, P_2, \ldots, P_k; P_{k+1}, \ldots, P_{2k+1}\}$ be an arbitrary set of $2k + 1$ points of M. (It is slightly more convenient to label the points this way than to use the notation of (1). Let $\tau(P_i)$, $1 \leq i \leq 2k + 1$ be the corresponding subsets of W. Let $\Omega = \{1, 2, 3, \ldots, 2k+1\}$ and let σ be a (possibly empty) subset of Ω. Put

$$r_\sigma = [\bigcap_{i \,\epsilon\, \sigma} \tau(P_i)] \cap [\bigcap_{j \,\not\epsilon\, \sigma} \overline{\tau(P_j)}]$$

where the bar denotes complementation. Thus, r_σ consists of those points of W belonging to $\tau(P_i)$ if $i \,\epsilon\, \sigma$ and not to $\tau(P_j)$ if $j \,\not\epsilon\, \sigma$. The sets r_σ are disjoint and their union is W. Put $m_\sigma = m(r_\sigma)$. Define

$$F(P_1, \ldots, P_{2k+1}) = \sum_{i=1}^{k} \sum_{j=k+1}^{2k+1} \rho(P_i, P_j) - \sum_{1 \le i < j \le k} \rho(P_i, P_j)$$

$$- \sum_{1 \le i < j \le k+1} \rho(P_i, P_j).$$

Then

$$(4) \quad F(P_1, \ldots, P_{2k+1}) = \sum_{i=1}^{k} \sum_{j=k+1}^{2k+1} m(\tau(P_i) \Delta \tau(P_j))$$

$$- \sum_{1 \le i < j \le k} m(\tau(P_i) \Delta \tau(P_j))$$

$$- \sum_{k+1 \le i < j \le 2k+1} m(\tau(P_i) \Delta \tau(P_j))$$

Now

$$\tau(P_i) \Delta \tau(P_j) = \left[\bigcup_{\substack{i \,\epsilon\, \sigma \\ j \,\not\epsilon\, \sigma}} r_\sigma\right] \cup \left[\bigcup_{\substack{i \,\not\epsilon\, \sigma \\ j \,\epsilon\, \sigma}} r_\sigma\right]$$

Since $\tau(P_i) \Delta \tau(P_j)$ consists of points belonging to $\tau(P_i)$ but not to $\tau(P_j)$ or to $\tau(P_j)$ but not to $\tau(P_i)$. Hence

$$(5) \quad m(\tau(P_i) \Delta \tau(P_j)) = \sum_{\substack{i \,\epsilon\, \sigma \\ j \,\not\epsilon\, \sigma}} m_\sigma + \sum_{\substack{i \,\not\epsilon\, \sigma \\ j \,\epsilon\, \sigma}} m_\sigma.$$

Let a_σ be the coefficient of m_σ after (5) is inserted in (4). We show $a_\sigma \ge 0$ for all subsets σ of Ω. This implies $F(P_1, P_2, \ldots, P_{2k+1}) \ge 0$, which is the desired conclusion (since k is arbitrary).

Suppose that σ contains α integers between 1 and k and β integers between $k + 1$ and $2k + 1$. The coefficient of m_σ in $\sum_{i=1}^{k} \sum_{j=k+1}^{2k+1} m(\tau(P_i) \Delta \tau(P_j))$ is $\alpha(k + 1 - \beta) + \beta(k - \alpha)$. The coefficient of m_σ in $\sum_{1 \leq i < j \leq k} m(\tau(P_i) \Delta \tau(P_j))$ is $\alpha(k - \alpha)$ and in

$\sum_{k+1 \leq i < j \leq 2k+1} m(\tau(P_i) \Delta \tau(P_j))$ is $\beta(k + 1 - \beta)$. Combining these, we

find $a_\sigma = (\alpha - \beta)^2 + (\alpha - \beta)$. As the function $g(x) = x^2 + x$ is

nonnegative for <u>integral</u> x, we have $a_\tau \geq 0$.

Now suppose that $M = 2^X$, the family of subsets of the finite

set, X. Let $P \subseteq X$, $Q \subseteq X$ and put $\rho(P,Q) = |P \Delta Q|$. Setting

$W = M$, $\tau(P) = P$, and $m(P) = |P|$, we infer at once from Theorem I

that (M,ρ) is hypermetric. In order for a metric space (M,ρ)

to be realizable as the distance space for a set of subsets of a

finite set, it is easily shown to be necessary that all triangles

of M have even perimeter. Tylkin [8] showed that this property

together with 2-hypermetricity is sufficient for realizability if

M has no more than 5 points. If $|M| \geq 6$, the problem of reali-

zability is more complicated, [3], [8], and has not been completely

solved.

Tylkin [8], was the first to formulate the inequality, (1),

chiefly in the context of the combinatorial problem of symmetric

difference cardinalities. He referred to (1) as the f-polygonal

inequality. Considerably later, the writer rediscovered (1), and

introduced the term "hypermetric", [4]. With the help of Theorem I

it became possible to demonstrate that the class of hypermetric

spaces is larger than was at first anticipated.

Geometric Examples

To see that the real line, R_1, equipped with the usual

metric, ρ, is hypermetric, we set $\tau(P) = [0,P]$ for each P in

R_1. It is immediate that $\rho(P,Q) = m(\tau(P) \Delta \tau(Q))$; here m denotes

Lebesgue measure on R_1.

Let $(S_{n,r}, \rho)$ be the n-sphere of radius r provided with the

great circle metric. With $P \in (S_{n,r}, \rho)$ let $\tau(P)$ be the solid

hemisphere centered at P. Let m denote volume in R_n. Then

$\rho(P,Q) = K_{n,r} m(\tau(P) \Delta \tau(Q))$ where $K_{n,r}$ is a constant depending only on n and r. This is immediate when $n = 1$ and $n = 2$ and can be established by a simple integration argument in the general case [5]. So Theorem I implies that $(S_{n,r}, \rho)$ is hypermetric.

Now let r tend to infinity and project from $S_{n,r}$ onto a tangent hyperplane isometric with R_n (n-dimensional Euclidean space). We see at once that R_n is hypermetric. (This way of proving that R_n is hypermetric was pointed out to me by Professor Ralph Alexander after the lecture. It is more efficient than the integration argument which appears in [4], since one wants to prove $(S_{n,r}, \rho)$ hypermetric in any case.

From the hypermetricity of $(S_{n,r}, \rho)$ one can quickly deduce an inequality for the sum of the mutual great-circle distances of m points on $S_{n,r}$ [5]:

(6) $\quad \sum_{1 \leq i < j \leq m} \rho(P_i, P_j) \leq \frac{m^2 - 1}{4} \pi r, \quad m$ odd

(7) $\quad \sum_{1 \leq i < j m} \rho(P_i, P_j) \leq \frac{m^2}{4} \pi r, \qquad m$ even.

These are best possible as one can see by taking $\frac{m-1}{2}$ (resp. $\frac{m}{2}$) points at the south pole.

Hypermetric Graphs

Let G be a connected undirected graph without loops or multiple edges. Let $\rho(P,Q)$ be the number of edges in the shortest path joining P and Q, if $P \neq Q$; put $\rho(P,P) = 0$. (G, ρ) is a metric space which may or may not be hypermetric.

Let K_n be the complete graph with n vertices. Then $K_n(\rho)$ is hypermetric. For we may assign a measure $\frac{1}{2}$ to each vertex of K_n and set $\tau(P) = P$, for all $P \epsilon K_n$. Then $\rho(P,Q) = m(\tau(P) \Delta \tau(Q))$ and the result follows from Theorem I.

Trees are hypermetric. To prove this, let T be a tree, and O an arbitrary vertex of T. Let $\tau(P)$ be the set of edges in

the unique path joining O to P. Let each edge have measure 1.
Then again $\rho(P,Q) = m(\tau(P) \Delta \tau(Q))$ and Theorem I may be applied.

On the other hand, circuits are also hypermetric. This may be
established by an argument which is the discrete analogue of the
proof of the hypermetricity of S_1 given in the previous section.

The graphs determined by the edges and vertices of the five
regular polyhedra are hypermetric. We give proofs for the
icosahedron and dedecahedron; similar and somewhat simpler proofs
can be constructed in the other three cases. Let I be the
icosahedral graph and let P be a vertex of I. Let $\tau(P)$ be the
set consisting of P and the five vertices adjacent to P. Let
each vertex of I have measure $\frac{1}{4}$. One can check by looking at
figure 1, that $\rho(P,Q) = m(\tau(P) \Delta \tau(Q))$. (Because of the symmetry,
there are only four cases to examine: $\rho(P,Q) = 0,1,2$ or 3.) In
like manner, if D is the dodecahedral graph and P is a vertex
of D, we can take as $\tau(P)$ the set of vertices at a distance
$0,1$ or 2 from P. Letting each vertex have measure $\frac{1}{4}$ and
looking at the figure, we can verify that $\rho(P,Q) = m(\tau(P) \Delta \tau(Q))$.
(Here we must examine the 6 cases $\rho(P,Q) = 0,1,2,3,4,5$.) The
reader will remark the close resemblance between these proofs and
the proof of the hypermetricity of the sphere. In each instance we
associated with P a "hemisphere", i.e., the set consisting of
that half of the vertex-set which is closest to P. (For the cube
and octahedron one uses faces rather than vertices, each face having
measure $\frac{1}{2}$ (resp. $\frac{1}{4}$).)

The bipartite graph $K_{3,2}$ is not hypermetric. Suppose that
the vertex set is $\{P_1,P_2,P_3;Q_1,Q_2\}$ where $\rho(P_i,P_j) = \rho(Q_i,Q_j) = 2$,
$i \neq j$ and $\rho(P_i,Q_j) = 1$. Then (1) is not satisfied; the left-hand
side is 8 and the right-hand side is 6. Consequently the bipartite graphs $K_{m,n}$, with $m \geq 3, n \geq 2$ are not hypermetric, for

they contain a subgraph isometric with $K_{3,2}$. (The metric on the subgraph is the same as that induced by the metric on $K_{m,n}$ in this case.) Since $K_{3,2}$ is planar, planar graphs need not be hypermetric. It would be valuable to be able to characterize hypermetric graphs by graph-theoretical properties.

The Four-Point Property

The space (M,ρ) is ultrametric if

$$(8) \quad \rho(P,Q) \leq \max(\rho(P,R),\rho(Q,R))$$

for all P,Q,R in M. A field metrized by a non-archimedean valuation is ultrametric. In [4] it is shown that every ultra-metric space is hypermetric.

A condition weaker than (8) is the four-point property:

$$(9) \quad \rho(P,Q) + \rho(R,S) \leq \max(\rho(P,R) + \rho(Q,S), \rho(P,S) + \rho(Q,R))$$

for all P,Q,R,S in M. Buneman [2] has shown that every tree possesses the four-point property and, further, that any connected graph without triangles that has a graphical distance with the four-point property must be a tree. Since trees are hypermetric, one may wonder whether every metric space with the four-point property is hypermetric. But Buneman has also shown in [2] that if ρ is a metric on a set S satisfying (9), then there is a weighted tree (a tree whose edges are assigned positive weights) which contains the members of S among its points and whose metric induces ρ. Now a weighted tree may be shown to be hypermetric by exactly the same argument that was used in the preceding section to prove that a tree is hypermetric. Thus we have

Theorem II: Any space with the four-point property is hyper-metric.

There is thus a chain of irreversible implications:

Ultrametric $\Rightarrow$ Four-Point $\Rightarrow$ Hypermetric $\Rightarrow$ Metric.

It would be well worthwhile to insert further links in this chain.

Lattices

Let $L = L(\vee, \wedge; <)$ be a lattice with operations $\vee, \wedge$ and order relation $<$. L is normed if a real, non-negative function $\| \ \|$ can be defined on L so that

(10) $\quad s < t, \ s \neq t \Rightarrow \|s\| < \|t\|$ and

(11) $\quad \|s \vee t\| + \|s \wedge t\| = \|s\| + \|t\|$

for each pair of elements (s,t) of L. A normed lattice can be made into a metric space by defining

(12) $\quad \rho(x,t) = \|s \vee t\| - \|s \wedge t\|$.

For example the set of finite subsets of a set X is a normed lattice with operations union and intersection, order relation inclusion and norm $\|s\| = |s|$. The metric defined by (12) is $\rho(s,t) = |s \ \Delta \ t|$.

It is well-known that every normed lattice is modular. On the other hand it is proved in [4] that a necessary and sufficient condition for $L(\vee, \wedge; <)$ to be distributive is that (L, ρ) be hypermetric. In fact, 2-hypermetricity is sufficient for distributivity. Thus the concept of hypermetricity enables us to give a purely metric characterization of distributive lattices among metric lattices.

Elliptic and Hyperbolic Spaces

The elliptic space $E_{n,r}$ is obtained from $S_{n,r}$ by identifying antipodal points. If distance in $S_{n,r}$ is denoted by $\bar{\rho}(x,y)$, the distance $\rho(x,y)$ is defined in $E_{n,r}$ by

$$\rho(x,y) = \bar{\rho}(x,y) \quad \text{if} \quad \bar{\rho}(x,y) \leq \frac{1}{2}\pi r$$

$$\bar{\rho}(x,y) = \pi r - \bar{\bar{\rho}}(x,y) \quad \text{if} \quad \bar{\bar{\rho}}(x,y) > \frac{1}{2} \pi r$$

Thus no two points of $E_{n,r}$ have distance greater than $\frac{1}{2} \pi r$. It can be shown [1, p.18], that $(E_{n,r}, \rho)$ is a metric space.

Clearly $E_{1,r}$ is isometric with $S_{1,r/2}$ and hence hypermetric. But, as the following example shows, $E_{2,r}$ is not even 2-hypermetric. Since $E_{2,r}$ can be isometrically embedded in $E_{n,r}$ if $n > 2$, the latter spaces are not hypermetric either.

Regard the points of $E_{2,r}$ as points of a hemisphere of radius r and center O as shown in Figure 2. Let S be the pole of the hemisphere. Let P_1, P_2, Q_1, Q_2 be points on a small circle of latitude $\pi/4$ and center O' so situated that the quadrilateral $P_1 Q_1 P_2 Q_2$ is a square. Q_3 is an intersection of the equator and the great circle joining P_1 and P_2. It is immediate that $\rho(P_1, P_2) = \rho(Q_1, Q_2) = \frac{\pi}{2} r$. Inasmuch as the Euclidean distances $O'P_1 = O'Q_1 = r/\sqrt{2}$, we have $P_1 Q = r$ and $P_1 O Q_1 = \pi/3$. Hence $\rho(P_1, Q_1) = \rho(P_1, Q_1) = \rho(P_2, Q_2) = \rho(P_2, Q_2) = \frac{\pi}{3} r$. Evidently $\rho(P_1, Q_3) = \frac{\pi}{4} r$, while $\bar{\rho}(P_2, Q_3) = \frac{3\pi}{4} r$ so that $\rho(P_2, Q_3) = \frac{\pi}{4} r$. Finally, since Q_3 is the pole of the great circle through Q_1 and Q_2, we have $\rho(Q_1, Q_3) = \rho(Q_2, Q_3) = \frac{\pi}{2} r$. Then $\rho(P_1, P_2) + \rho(Q_1, Q_2) + \rho(Q_1, Q_3) + \rho(Q_2, Q_3) = 2\pi r$ while $\rho(P_1, Q_1) + \rho(P_1, Q_2) + \rho(P_1, Q_3) + \rho(P_2, Q_1) + \rho(P_2, Q_2) + \rho(P_2, Q_3) = 11 \frac{\pi r}{6}$ contradicting (1) with $k = 2$. Thus $E_{2,r}$ is not 2-hypermetric. In [9], the stronger result that $E_{2,r}$ is not even quasi-hypermetric (see below) is established.

The hyperbolic space $H_{n,r}$ is defined in [1]. $H_{2,r}$ provides a model for Lobatchevskian geometry. It can easily be shown that $H_{1,r}$ is hypermetric, but for $n \geq 2$, the problem of the hypermetricity of $H_{n,r}$ is open.

Quasi-Hypermetric Spaces

If the semi-metric space (M, ρ) satisfies the condition (3) for all choices of $\{P_1, P_2, \ldots, P_n\} \subseteq M$ where the quantities $x_1, x_2, \ldots, x_n$ are <u>real</u> numbers with

$$(13) \quad \sum_{i=1}^{n} x_i = 0,$$

then (M, ρ) is said to be quasi-hypermetric. It is shown in [6] that every hypermetric space is quasi-hypermetric. The converse is false; indeed, a quasi-hypermetric space need not even be metric. As simple examples show, it is not useful to replace conditions (2) or (13) by $\sum_{i=1}^{n} x_i = a$, where $a \neq 0$ or 1.

Normed Linear Spaces and Zonoids

Let M be a real linear space with norm $\| \ \|$ and metric $\rho(P, Q) = \|P - Q\|$. Such a space is not necessarily hypermetric; as shown in [4], the space R_3 with $\|(x_1, x_2, x_3)\| = \max(|x_1|, |x_2|, |x_3|)$ is not even 2-hypermetric.

A simple theorem, [4], enables one to construct hypermetric spaces by taking direct products.

<u>Theorem III</u>. Let $(M_1, \rho_1), (M_2, \rho_2), \ldots, (M_h, \rho_h)$ be k-hypermetric. Let $M = M_1 \times M_2 \times \ldots \times M_h$ be metrized by

$$\rho(P_1, \ldots, P_h), (Q_1, \ldots, Q_h)) = \sum_{i=1}^{h} \rho_i(P_i, Q_i)$$

Then (M, ρ) is k-hypermetric.

As usual, let $L(p, n)$ denote R_n with norm $\|x_1, x_2, \ldots, x_n\| = (\sum_{i=1}^{n} |x_i|^p)^{1/p}$; let $L(\infty, n)$ be R_n with norm $\|x_1, x_2, \ldots, x_n\| = \max(x_1, \ldots, x_n)$. The above example shows that $L(\infty, 3)$ is not 2-hypermetric and also that $L(\infty, n)$, $n \geq 3$ is not 2-hypermetric. Since $\lim_{p \to \infty} (\sum_{i=1}^{n} |x_i|^p)^{1/p} = \max(|x_1|, |x_2|, \ldots, |x_n|)$ it follows that $L(p, n)$ is not 2-hypermetric if $n \geq 3$ and p is sufficiently large. On the other hand $L(2, n)$ is hypermetric since it is simply E_n. $L(1, n)$ is also hypermetric, in virtue of Theorem III.

Another familiar class of normed linear spaces is the class $L_p(0,1)$ of pth power integrable functions with

$$\|f\| = \left[\int_0^1 |f(x)|^p \, dx \right]^{1/p}.$$

$L_2(0,1)$ is hypermetric because it is an inner product space so that any finite subset is isometric with a subset of some Euclidean space. The argument used to prove Theorem III may be extended to show that $L_1(0,1)$ is hypermetric. Lindenstrauss [7] has shown that any normed linear space of dimension ≤ 2 may be isometrically embedded in $L_1(0,1)$. Thus normed linear spaces of dimension ≤ 2 are hypermetric. So our example above of a 3-dimensional non-hypermetric normed linear space is about as simple as possible.

Witsenhausen [9] has demonstrated that every quasi-hypermetric normed linear space is hypermetric. His argument also establishes points of contact with the <u>zonoid problem</u>. A zonoid is a closed, centrally symmetric convex body that can be approximated arbitrarily closely by zonotopes. A zonotope is a polytope, Z, which is generated by a finite number, N, of vectors v_i in R_n by translation, that is,

$$Z = \{ \sum_{i=1}^n \lambda_i v_i \,|\, 0 \leq \lambda_i \leq 1, | \leq i \leq N \}.$$

It is not always easy to decide whether a given body is a zonoid. Witsenhausen shows that if the finite-dimensional normed linear space M is quasi-hypermetric, the unit ball of M^*, the dual of M, is a zonoid; he then shows that if the latter is true, M is hypermetric. Hence the problem of deciding whether a finite-dimensional space is hypermetric or quasi-hypermetric is equivalent to the zonoid problem. This observation may facilitate a numerical approach to the zonoid problem well adapted to the use of computers. The reader desiring further information about zonoids should consult the excellent articles by E. Bolker: [10], [11].

Another important result of Witsenhausen is that $L_p(0,1)$, and, a fortiori, $L(n,p)$ is hypermetric if $1 \leq p \leq 2$. In addition, letting p_n denote the smallest p such that the unit ball of $L(n,p)$ is a zonoid, he shows that $p_3 \geq \log 3/\log 2$ and $p_n \geq 2 - 1/2n \log 2 + o(n^{-1})$, substantially improving previous bounds. His proof proceeds by an examination of the quasi-hypermetric inequality in the dual space. It follows at once that $L_p(0,1)$ is not hypermetric for $p > 2$, that $L(3,p)$ is not hypermetric if $p > \log 3 | (\log 3 - \log 2) = 2.709...$ and that $L(p,n)$ is not hypermetric if $p > 2 + 1/2n \log 2 + o(n^{-1})$. Intermediate cases are still open.

Cases of Equality

A close examination of the proof of Theorem I will yield information about the occurrence of equality in (1). It is convenient first to define sets u_j^* and v^* as follows: $u_j^*(P_1, \ldots, P_k)$ is the set of all elements of W belonging to at least j of the sets $\tau(P_1), \ldots, \tau(P_k)$ while $v_j^*(Q_1, \ldots, Q_{k+1})$ is the set of all elements of W belonging to at least j of the sets $\tau(Q_1), \ldots, \tau(Q_{k+1})$. Then it is not hard to prove

Theorem IV: Under the hypotheses of Theorem I, we have

$$\sum_{1 \leq i < j \leq k} \rho(P_i, P_j) + \sum_{1 \leq i < j \leq k+1} \rho(Q_i, Q_j) = \sum_{i=1}^{k} \sum_{j=1}^{k+1} \rho(P_i, Q_j)$$

if and only if the collections $\{u_j^*\}$ and $\{v_j^*\}$ almost separate each other, that is

$$v_{k+1}^* \subseteq u_k^* \subseteq v_k^* \subseteq \cdots \subseteq v_{j+1}^* \subseteq u_j^* \subseteq v_j^* \cdots \subseteq u_1^* \subseteq v_1^*$$

except for sets of measure zero.

From Theorem IV we can readily deduce that on E_1, equality holds in (1) if and only if the sets of points $\{P_i\}$ and $\{Q_i\}$

separate each other; that is, with some labelling of the points we have

$$Q_1 \leq P_1 \leq Q_2 \leq P_2 \leq \cdots \leq Q_j \leq P_j \leq Q_{j+1} \leq \cdots \leq P_k \leq Q_{k+1}$$

It is possible to show, with greater difficulty, [4], that equality holds in E_n if and only if some m, $0 \leq m \leq k$ of the points $P_1, \ldots, P_k$ are pairwise equal to some m of the points $Q_1, \ldots, Q_{k+1}$ while the remaining points lie on a single straight line with the sets of P-points and Q-points separating each other.

REFERENCES

[1] L. Blumenthal, Theory and Applications of Distance Geometry,
     Oxford University Press, London, 1953.

[2] P. Buneman, A Note on the Metric Properties of Trees, J. Comb.
     Theory, Series B 17 (1974), 48-50.

[3] J.B. Kelly, Products of Zero-One Matrices, Can. J. Math. 20
     (1968), 298-329.

[4] J.B. Kelly, Metric Inequalities and Symmetric Difference In-
     equalities II - edited by O. Shisha, Academic Press (1970),
     193-212.

[5] J.B. Kelly, Combinatorial Inequalities - Combinatorial Struc-
     tures and their Applications - edited by R. Guy, H. Hanani,
     N. Sauer and J. Schonheim, Gordon and Breach, New York, 1970,
     201-208.

[6] J.B. Kelly, Hypermetric Spaces and Metric Transforms - Inequal-
     ities II - edited by O. Shisha, Academic Press, New York
     (1972), 149-158.

[7] J. Lindenstrauss, On extension of operators with finite range,
     Illinois J. Math 8 (1964), 488-499.

[8] M.E. Tylkin, On Hamming Geometry of Unitary Cubes (Russian)
     Doklady Akad. Nauk SSSR 134 No. 5 (1960), 1037-1040.

[9] H.S. Witsenhausen, Metric Inequalities and the Zonoid Problem,
     Proc. A.M.S. 40 (2), 1973, 517-520.

[10] E.D. Bolker, A Class of Convex Bodies, Trans. Am. Math. Soc.
     145 (1969), 323-345.

[11] E.D. Bolker, The Zonoid Problem, Am. Math. Monthly 78 (1971),
     529-531.

METRIC PROBLEMS IN ELLIPTIC GEOMETRY

Johan J. Seidel
Technological University
The Netherlands

1. Introduction

More than 25 years ago, in different parts of the world, 2 young students worked on the same problem. In Missouri L.M. Kelly [7], our present host, was guided by Professor L.M. Blumenthal, whom we honor in this conference. In Holland the late Professor J. Haantjes suggested to the present author [13] the very same problem: to determine the congruence order of the elliptic plane. It seems worthwhile to review the problem, its solutions and the progress which has been made for the general case in these 25 years. Unfortunately, today the congruence order of the elliptic space E^n is still unknown for $n > 2$. On the other hand, we are now able to explain the reasons for the difficulty of the problem, and at least can present a reasonable conjecture. It became clear that our problem is related to a great variety of other subjects, both of pure and applied nature. As such we mention finite simple groups, spherical harmonics, error correcting codes, electrical networks, and statistical designs. The past 25 years have shown great progress in these fields. For instance, apart from the 5 known Mathieu groups, now 18 new sporadic simple groups have been discovered [3]; algebraic coding theory has developed from rudiments to a mature field [11].

In the present paper we wish to report on several problems which may be interpreted as metric problems in elliptic geometry. We indicate some of the difficulties which occur. The paper may serve as an illustration to the remark by Professor Blumenthal ([1], p. 269) concerning the state of affairs in 1953 of the elliptic space problem:

"...that is the best that can be said at this time. More powerful methods must be developed before the general problem can be approached".

2. Congruence Order

Any metric space M is isometrically imbeddable in the Euclidean plane whenever each 5 element subset of M is isometrically imbeddable in the Euclidean plane. This has been proved by Menger [12]. He calls the number 5 (which cannot be reduced to 4) the congruence order of the Euclidean plane. In addition, Menger proved that any M is isometrically imbeddable in Euclidean n-space whenever each (n + 3)-subset of M is. The number n + 3 is the smallest number having this property, and is called the congruence order of Euclidean n-space. Analogously, Blumenthal [1] has determined the congruence order of hyperbolic n-space, and of spherical n-space. Both numbers trun out to be n + 3.

For elliptic space the situation is completely different. The congruence order of the elliptic plane turns out [6] to be 7 and not 5, and that of elliptic (n - 1)-space, for n > 3, is unknown but much larger than n + 3. In order to explain this different behavior we first recall some facts from elliptic geometry.

3. Elliptic Space E^{n-1}

In the vector space $\mathbb{R}^n$ of dimension n over the reals we call elliptic points the lines through the origin O, elliptic lines the planes through O, elliptic planes the 3-subspaces, etc. The elliptic distance between 2 elliptic points is defined to be the angle between the corresponding lines. Notice, that all distances are at most $\pi/2$. Thus elliptic (n - 1)-space E^{n-1} is defined in terms of $\mathbb{R}^n$.

The 6 diagonals of an icosahedron in $\mathbb{R}^3$ yield 6 elliptic points in the elliptic plane E^2 whose elliptic distances are all equal. In fact, these distances are arccos $1/\sqrt{5}$. This explains why the congruence order E^2 cannot be 6. Indeed, let the metric space M consist of 100 points all of whose distances are arccos $1/\sqrt{5}$. Then each 6-subset of M is isometrically imbeddable in E^2, but M is not.

For the congruence order in E^{n-1} we should at least know the maximum number of elliptic points all of whose distances are equal. We shall denote this number by $N(n)$, the maximum number of lines in $\mathbb{R}^n$ all of whose angles are equal. As a result of investigations in recent years, $\lceil 10 \rceil$, $\lceil 8 \rceil$, $\lceil 16 \rceil$, the following values for $N(n)$ are known.

$\quad$ n $\quad = 2,3,4, 5, 6, 7,\ldots,13,15, 21, 22, 23,$

$\quad$ $N(n) = 3,6,6,10,16,28,\ldots,28,36,126,176,276,$

$\quad 28 \leq N(14) \leq 30, \ 40 \leq N(16) \leq 42, \ 48 \leq N(17) \leq 51,$

$\quad 72 \leq N(19) \leq 76, \ 90 \leq N(20) \leq 96, \ 344 \leq N(43).$

For $n \geq 6$ we have $\lceil 8 \rceil$

$$n\sqrt{n} \leq N(n) \leq \frac{1}{2} n(n + 1).$$

Here is an (easy) conjecture, which holds true for $n = 2,3$.

$\quad$ <u>Conjecture</u>. The congruence order of E^{n-1} equals $N(n) + 1$.

4. <u>Pillars</u>

$\quad$ Going into more detail we shall now explain the explosion of $N(n)$ for $n = 7$, and the constancy for $7 \leq n \leq 13$. The explosion for $n = 23$ (and the constancy for $23 \leq n \leq 41$) are considered in $\lceil 8 \rceil$ along similar lines.

$\quad$ <u>Definition</u>. $N_{\frac{1}{3}}(n)$ is the maximum number of lines in $\mathbb{R}^n$ all of whose angles equal arccos $\frac{1}{3}$.

$\quad$ The first part of the following theorem proves an (adapted) conjecture by Blumenthal and Kelly [2], which turns out to be true for almost all n.

$\quad$ <u>Theorem</u>. $N_{\frac{1}{3}}(n) = 2(n - 1)$, for $n = 2,3,4,$ and $n \geq 15,$ and

$$n \qquad = 3,4,\ 5,\ 6,\ 7,\dots,15,$$

$$N_{\frac{1}{3}}(n) \ = 4,6,10,16,28,\dots,28.$$

For the proof we first remark that the existence in $\mathbb{R}^n$ of a set of lines all of whose angles are arccos $\frac{1}{3}$, is equivalent to the existence in $\mathbb{R}^n$ of a set, of the same cardinality, of unit vectors all of whose inner products are $\pm\frac{1}{3}$. Indeed, any line through the origin is determined by a unit vector spanning the line, and the angle between 2 lines equals the angle, or the supplement of the angle, between their unit vectors. In $\mathbb{R}^3$ the unit vectors p_1, p_2, p_3, p_4 of the figure, from the center of a cube of size $2/\sqrt{3}$ to 4 of its vertices, have the inner products $(p_i, p_j) = -\frac{1}{3}$, $i \neq j = 1,2,3,4$.

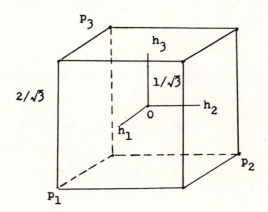

It is easily seen that these vectors establish $N_{\frac{1}{3}}(3) = 4$. For the maximum set in $\mathbb{R}^n$ we start, without loss of generality, with these vectors p_1, p_2, p_3, p_4 and look for further vectors $x \in \mathbb{R}^n$ such that

$$(x,x) = 1, \quad (x,p_i) = \pm\frac{1}{3} =: \frac{1}{3}\,\varepsilon_i, \quad i = 1,2,3,4.$$

Now $p_1 + p_2 + p_3 + p_4 = 0$ implies $\varepsilon_1 + \varepsilon_2 + \varepsilon_3 + \varepsilon_4 = 0$, whence $(\varepsilon_1, \varepsilon_2, \varepsilon_3, \varepsilon_4) = (1,-1,-1,1)$ or $(-1,1,-1,1)$ or $(-1,-1,1,1)$.

Decomposing

$$x = h + c, \quad h \in \mathbb{R}^3, \quad c \perp \mathbb{R}^3,$$

we have $(h, p_i) = \frac{1}{3} \varepsilon_i$, $i = 1, 2, 3, 4$. Hence the following possibilities are available for h:

$$h_1 := \frac{1}{4}(p_1 - p_2 - p_3 + p_4), \quad h_2 := \frac{1}{4}(-p_1 + p_2 - p_3 + p_4),$$

$$h_3 := \frac{1}{4}(-p_1 - p_2 + p_3 + p_4).$$

Thus, for x we must look in the 3 linear varieties of dimension $n - 3$, which intersect the basic $\mathbb{R}^3$ perpendicularly in h_1, h_2, h_3, respectively. These are called the 3 <u>pillars</u>, erected in h_1, h_2, h_3 perpendicular to $\mathbb{R}^3$. For 2 such vectors $x = h + c$ and $x' = h' + c'$ we have

$$(x, x') = (h, h') + (c, c').$$

Let the $N \times N$ matrix x denote the Gramian matrix of the inner products of N such vectors x, and let H and C denote the Gramian of the corresponding h and c, respectively. Then

$$X = H + C,$$

with positive semidefinite C, of rank $\leq n - 3$.

Now let us investigate how many such vectors x can take place in one pillar. In this case the matrix equation just mentioned reads

$$I + \frac{1}{3} A = \frac{1}{3} J + C,$$

where J is the $N \times N$ all-one matrix, and A is an $N \times N$ symmetric matrix with elements 0 on the diagonal and ± 1 elsewhere. The positive semidefinite matrix

$$C = \frac{2}{3} I + \frac{1}{3} (A - J + I)$$

is the Gramian of N vectors in $(n - 3)$-space at angles 90° or
180°, since their cosines are 0 or -1. Obivously, the maximum number of such vectors is $2(n - 3)$, which also is the maximum possible
number of vectors x in any one pillar.

Finally, let us investigate the case that the pillar on h_1
contains one vector, and the pillar on h_2 contains an orthogonal
set of s vectors. Then the equation for the $(s + 1) \times (s + 1)$
matrices reads

$$
I + \frac{1}{3}
\begin{bmatrix}
0 & a_{12} & \cdots & a_{1s} \\
a_{12} & & & \\
\vdots & & J_s - I_s & \\
\vdots & & & \\
a_{1s} & & &
\end{bmatrix}
= \frac{1}{3}
\begin{bmatrix}
1 & 0 & \cdots & 0 \\
0 & & & \\
\vdots & & J_s & \\
\vdots & & & \\
0 & & &
\end{bmatrix}
+ C,
$$

$$
\frac{3}{2} C =
\begin{bmatrix}
1 & \frac{1}{2}a_{12} & \cdots & \frac{1}{2}a_{1s} \\
\frac{1}{2}a_{12} & & & \\
\vdots & & I_s & \\
\vdots & & & \\
\frac{1}{2}a_{1s} & & &
\end{bmatrix}.
$$

Since $\det C \geq 0$, this implies

$$
1 - \frac{1}{4}(a_{12}^2 + \ldots + a_{1s}^2) = 1 - \frac{1}{4}s \geq 0
$$

and we have $s \leq 4$, _independent of the dimension_ n.

For $n = 7$ the 3 pillars may be filled by 8 vectors each,
which implies

$$
N_{\frac{1}{3}}(7) \leq 4 + 3 \times 8 = 28.
$$

On the other hand, 28 such vectors are easily constructed, whence
$N_{\frac{1}{3}}(7) = 28$. Indeed, consider the $(0,1)$ incidence matrix of the

Fano plane PG(2,2). From this 7 x 7 matrix an 28 x 7 matrix is constructed by replacing, in each row, the 3 ones by the columns of

$$
\begin{bmatrix}
1 & 1 & 1 \\
1 & -1 & -1 \\
-1 & 1 & -1 \\
-1 & -1 & 1
\end{bmatrix} ,
$$

and the 4 zeros by zero columns of size 4. The 28 rows of the resulting matrix define 28 vectors in $\mathbb{R}^7$ having inner products ± 1, and cosines $\pm \frac{1}{3}$. For $n > 7$ the contribution of the 3 pillars cannot exceed 3×8, unless 2 pillars are empty and the remaining pillar contains $2(n - 3) > 24$ vectors. This happens for $n > 15$, and then

$$
N_{\frac{1}{3}}(n) = 4 + 2(n - 3) = 2(n - 1).
$$

The remaining cases

$$
N_{\frac{1}{3}}(6) = 4 + 3.4, \quad N_{\frac{1}{3}}(5) = 4 + 3.2
$$

need some further reasoning, and may provide an exercise for the reader.

5. Further Results

We have seen that the congruence order problem leads naturally to the investigation of equidistant point sets in elliptic space E^{n-1}. Further questions of this metric nature may be posed. For instance, which 2-distance point sets may occur in E^{n-1}, and which 3-distance point sets? In the next section we shall see that maximal such sets bear relevance to the theory of finite simple groups.

Jointly with P. Delsarte and J.M. Goethals, the author announces [4] some results on bounds for the cardinality of sets of points in E^{n-1} whose distance take a restricted number of values. Let D be a finite subset of the real interval $[0,1[$, and let X be a finite subset of E^{n-1} having the property that

$$\cos^2 d(x,y) \in D, \quad \text{for all} \quad x \neq y \in X.$$

There are 2 types of bounds for card X. The special bound uses the values of the admitted distances. The absolute bound uses the number of such distances, not their values.

D	special bound	absolute bound
$\{\alpha\}$	$\dfrac{n(1-\alpha)}{1-n\alpha}$	$\begin{bmatrix} n+1 \\ 2 \end{bmatrix}$
$\{0,\alpha\}$	$\dfrac{n(n+2)(1-\alpha)}{3-(n+2)\alpha}$	$\begin{bmatrix} n+2 \\ 3 \end{bmatrix}$
$\{\alpha,\beta\}$	$\dfrac{n(n+2)(1-\alpha)(1-\beta)}{3-(n+2)(\alpha+\beta)+n(n+2)\alpha\beta}$	$\begin{bmatrix} n+3 \\ 4 \end{bmatrix}$
$\{0,\alpha,\beta\}$	$\dfrac{n(n+2)(n+4)(1-\alpha)(1-\beta)}{15-3(n+4)(\alpha+\beta)+(n+2)(n+4)\alpha\beta}$	$\begin{bmatrix} n+4 \\ 5 \end{bmatrix}$

The special bounds hold under certain restriction, e.g. the positivity of the denominators. The bounds are obtained by use of matrix techniques and the addition formula for Jacobi polynomials. Maximal sets with respect to these bounds sometimes provide combinatorial configurations with an interesting algebraic structure. On the other hand, the bounds need not be best possible. For instance, for the elliptic plane the maximal set for the distance 60° and 90° is realized by the 6 diagonals of the cubeoctahedron, but 6 does not meet either bound.

A further generalization of equidistant point sets in elliptic space is provided by the following. Recall that 2 subspaces Γ and Δ of E^{n-1}, of the same dimension $m-1$, are <u>Clifford parallel</u> whenever the distance d from any point $P \in \Gamma$ to Δ is independent of the choice of $P \in \Gamma$. So d depends on Γ and Δ only. In [9] the following problem is proposed.

Find the maximum number of $(m-1)$-subspaces in E^{n-1} which are pairwise Clifford parallel with the same distance d.

For m = 1 we are back in the case of equidistant point sets. For n = 4, m = 2, there are 4, and no more lines in elliptic 3-space which are pairwise Clifford parallel at distance arccos $1/\sqrt{3}$. For the general case, certain bounds have been derived. In addition, the complete solution for the case n = 2m has been established [9], by use of the Hurwitz-Radon matrix equations [17].

6. Finite Groups, Error-Correcting Codes, and Two-Graphs

The symmetry group of an equidistant point set X in elliptic space E^{n-1} is the group of all elliptic transformations which leave the set X invariant. It is interesting to observe that sometimes the symmetry group of a maximal equidistant point set X, in the sense of section 3, acts 2-transitively on X and is isomorphic to a simple group. We mention the following examples [8], [16], [14]: N(3) = 6 with the alternating group A_5; N(7) = 28 and N(15) = 36 with the symplectic group Sp(6,2): N(21) = 126 with the unitary group $P\Gamma U(3,5^2)$; N(22) = 176 with the Higman Sims group; N(23) = 276 with the Conway group ·3.

Conway's group ·O is the symmetry group of the Leech lattice [3], that is, the unique lattice Γ over the integers in $\mathbb{R}^{24}$ such that

$$\det \Gamma = 1, \quad \forall_{x \in \Gamma} ((x,x) \equiv 0 \bmod 2), \quad \min_{0 \neq x \in \Gamma} (x,x) \geq 4.$$

Consider vectors x and y of minimal norm (x,x) = (y,y) = 4. Then

$$4 \leq (x - y, x - y) = (x,x) + (y,y) - 2(x,y),$$

whence $|(x,y)| \leq 2$. For the angle δ between x and y we have

$$\cos \delta (x,y) \in \{0, \pm \tfrac{1}{4}, \pm \tfrac{1}{2}\}.$$

This yields a 3-distance point set in E^{23} of cardinality $\binom{28}{5}$, meeting the bounds announced in section 5. The symmetry group of this set is Conway's group ·1. It's subgroup .2 is the symmetry group of $2300 = \binom{25}{3}$ points in E^{22} with $\cos d \in \{0, \frac{1}{3}\}$, and the symmetry group of 2300 equidistant points in E^{275} with $\cos d = \frac{1}{17}$.

Among the subgroups of Conway's group ·O are the Mathieu groups $M_{11}, M_{12}, M_{22}, M_{23}, M_{24}$, which also act as symmetry groups of few-distance point sets in elliptic space. We describe one of these sets, just to indicate the relations between elliptic geometry and certain error-correcting codes. The code in question is the extended Golay code, which contains 2^{12} code words and is 3-error-correcting. For our elliptic purposes we define the code as follows (cf. [5]). Consider the 12 rows of the 12 x 24 matrix

$$\begin{bmatrix} 1 & j^T & -1 & -j^T \\ -j & J-2I & -j & Q \end{bmatrix},$$

where I is the unit matrix and j is the all-one vector of size 11, and

$$Q = \text{circul}(1,-1,1,-1,-1,-1,1,1,1,-1,1).$$

From any 2 rows of the matrix a new vector of length 24 is constructed by coordinate-wise multiplication. From [5], p. 613, it follows that the 2×2^{22} vector in $\mathbb{R}^{24}$, thus obtained, have length $\sqrt{24}$ and inner products $\in \{0, 8, -8\}$. This yields a 2-distance point set in E^{23} of cardinality 2^{11}, with $\cos d \in \{0, \frac{1}{3}\}$.

Finally, let us return to equidistant point sets in elliptic space, and mention an abstract characterization of such sets. In order to provide a combinatorial setting for the 2-transitive representation of certain sporadic groups, G. Higman proposed the notion of regular two-graphs, [14], [16], [15]. Recall that an ordinary graph is a pair of a set Ω and a map $f: \Omega^{(2)} \to \{1,-1\}$. Here $\Omega^{(k)}$ denotes the set of all unordered k-subsets of Ω.

Definition. A two-graph is a pair of a set Ω and a map $g: \Omega^{(3)} \rightarrow \{1, -1\}$ such that, for each $\{\alpha, \beta, \gamma, \delta\} \in \Omega^{(4)}$,

$$g\{\alpha, \beta, \gamma\} g\{\alpha, \beta, \delta\} g\{\alpha, \gamma, \delta\} g\{\beta, \gamma, \delta\} = 1.$$

Any equidistant point set in elliptic space constitutes a two-graph. Indeed, there are 2 types of equidistant triples of elliptic points at distance d: those spanned by 3 vectors at angles arccos d, and those spanned by 3 vectors at angles π-arccos d. Furthermore, any quadruple of elliptic points contains even number of triples of either type. Hence we have a two-graph. Conversely, let any two-graph (Ω, g) be given. We select any $p \in \Omega$ and define the symmetric matrix A on $\Omega \times \Omega$ by its elements $A(p,p) = A(i,i) = 0$, $A = (p,i) = A(i,p) = 1$, $A(i,j) = A(j,i) = g\{p,i,j\}$, for all $i \neq j \in \Omega \{p\}$. This definition is consistent. Let $-\lambda$ be the smallest eigenvalue of A, let $|\Omega| - n$ be its multiplicity, and suppose $\lambda \neq 1$. The matrix $A + \lambda I$, which is symmetric and positive semi-definite of rank n, is the Gramian of $|\Omega|$ vectors in $\mathbb{R}^n$ of length $\sqrt{\lambda}$, with inner products ± 1. These yield $|\Omega|$ elliptic points in E^{n-1} at mutual distance arccos $1/\lambda$. Now we essentially have proved the following theorem, cf. $\lceil 14 \rceil$, $\lceil 16 \rceil$.

Theorem. There is a one-to-one correspondence between the two-graph on t vertices, and the dependent sets of t equidistant elliptic points.

REFERENCES

[1]   L.M. Blumenthal, Theory and applications of distance geome-
        try, Oxford University Press (1953).

[2]   L.M. Blumenthal, L.M. Kelly, New metric-theoretic properties
        of elliptic space, Revista Univ. Nac. Tucumán, A 7 (1949),
        81-107.

[3]   J.H. Conway, Three lectures on exceptional groups, Chapter 7
        of "Finite simple groups", ed. M.B. Powell and G. Higman,
        Academic Press (1971), 215-247.

[4]   P. Delsarte, J.M. Goethals, J.J. Seidel, Bounds for systems
        of lines, and Jacobi polynomials, Philips Research Reports,
        to be published.

[5]   J.M. Goethals, J.J. Seidel, Strongly regular graphs derived
        from combinatorial designs, Canad. J. Math. 22 (1970), 597-
        614.

[6]   J. Haantjes, J.J. Seidel, The congruence order of the ellip-
        tic plane, Proc. Kon. Ned. Akad. Wetensch. A 50 (1948), 892-
        894 = Indag. Math 9 (1947), 403-405.

[7]   L.M. Kelly,
        Univ. Missouri, thesis (1948).

[8]   P.W.H. Lemmens, J.J. Seidel, Equiangular lines, J. Algebra 24
        (1973), 494-512.

[9]   P.W.H. Lemmens, J.J. Seidel, Equi-isoclinic subspaces of
        Euclidean spaces, Proc. Kon. Ned. Akad. Wetensch. A 76 (=
        Indag. Math. 35 (1973), 98-107.

[10]   J.H. van Lint, J.J. Seidel, Equilateral point sets in ellip-
        tic geometry, Proc. Kon. Ned. Akad. Wetensch. A 69 (= Indag.
        Math. 28) (1966), 335-348.

[11]   J.H. van Lint, Coding Theory, Lecture notes in mathematics
        201, Springer (1971).

[12]   K. Menger, Untersuchunger über allgemeine Metrik, Math. Ann.
        100 (1928), 75-163.

[13]   J.J. Seidel, De congruentie-orde van het elliptische vlak,
        Universiteit van Leiden, thesis (1948), 75 pp.

[14]   J.J. Seidel, A survey of two-graphs, Proc. Int. Coll. Teorie
        Combinatorie, Acc. Naz. Lincei, Roma, to appear.

[15]   J.J. Seidel, Graphs and two-graphs, 5th Southeastern Confer-
        ence on Combinatorics, Graph Theory, and Computing, to appear.

[16]   D.E. Taylor, Regular two-graphs, Proc. London Math. Soc., to
        appear.

[17]   Y.C. Wong, Isoclinic n-planes in Euclidean 2n-space, Clifford
        parallels in elliptic (2n - 1)-space, and the Hurwitz matrix
        equations, Memoir Amer. Math. Soc. 41 (1961).

DISCREPANCY AND SUMS OF DISTANCES BETWEEN POINTS OF A METRIC SPACE

Kenneth B. Stolarsky
University of Illinois

ABSTRACT

The following problem has already been studied: how should N points $p_1, \ldots, p_N$ be placed on the surface of the unit sphere of n-dimensional Euclidean space so that the sum $S(N; p_1, \ldots, p_N)$ of all $\binom{N}{2}$ distances which they determine is maximal? There is a "natural" discrepancy $D(p_1, \ldots, p_N)$ which measures the extent to which these points fail to be "uniformly distributed", and such that $S(N; p_1, \ldots, p_N) + D(p_1, \ldots, p_N)$ is constant. Here an analogous result is proved for metric spaces arising from sets on which a non-negative measure is defined. (More precisely, we associate with such a set S the metric space consisting of all the measure of their symmetric difference). This result is reminiscent of the combinatorial foundations of John B. Kelly's theory of hypermetric spaces.

Let the real interval $I = [a, b]$ have density $w(x)$ at x. Let $I_1, \ldots, I_N$ be subintervals of I, and denote the sum of the lengths of the I_i by L. The above result is applied to estimate the total mass of the $\binom{N}{2}$ intervals $I_i \cap I_j$ in terms of L, the masses of the I_i, and the function $w(x)$. If $m(J)$ denotes the mass of the interval J,

DISCREPANCY AND SUMS OF DISTANCES BETWEEN POINTS OF A METRIC SPACE

Kenneth B. Stolarsky
University of Illinois

§1. Introduction.

Let K be a bounded subset of a metric space M and let $p_1, \ldots, p_N$ be points of K. Let $S(N; p_1, \ldots, p_N)$ denote the sum of all distances determined by these points. How should one distribute the points p_i so that S is as large as possible? In [6] it was shown that if U is the surface of the unit sphere in n-dimensional Euclidean space, where $n \geq 2$, then the points should be distributed as uniformly as possible. More precisely, it is shown that (i) there is a natural way to define a nonnegative real number $D(p_1, \ldots, p_N)$, called the "discrepancy of $p_1, \ldots, p_N$", which measures the extent to which these points fail to be uniformly distributed, and that (ii) the sum

$$S(N; p_1, \ldots, p_N) + D(p_1, \ldots, p_N)$$

is constant; see [6, equation (2.4)]. The main result of this paper, equation (3.1), is an analogous formula for certain metric spaces arising from measures. Actually, equation (5.6) is much more satisfactory than equation (3.1), which it includes as a special case. However, both to avoid too much generality at the start, and also because it is convenient to use a consequence of (3.1) in the proof of (5.6), we do not proceed directly to (5.6).

As an application of our formula we shall estimate the amount of "overlap" in a system of real intervals $I_1, \ldots, I_N$, each of which is contained in a fixed real interval $I = [a, b]$. If λ denotes Lebesgue measure (i.e. length) and $\sum_{i=1}^{N} \lambda(I_i) = L$, it is not hard to show that when $I = [0, 1]$ we have

$$(1.1) \quad \sum_{i<j} \lambda(I_i \cap I_j) \geq \binom{L}{2}.$$

Let $w(x)$ denote a (weight) function which is positive and continuous on the closed interval $I = [a,b]$. For $T \subseteq I$ define

$$m(T) = \int_T w(x) d\lambda(x).$$

Let h be the harmonic mean of $w(x)$; in other words,

$$h \int_I \frac{d\lambda(x)}{w(x)} = \lambda(I).$$

Then we can show

$$(1.2) \quad \sum_{i<j} m(I_i \cap I_j) \geq \frac{1}{2}\{\frac{L^2 h}{b-a} - \sum_{i=1}^{N} m(I_i)\}$$

where

$$L = \sum_{i=1}^{N} \lambda(I_i).$$

For $w(x) \equiv 1$ and $[a,b] = [0,1]$ this reduces to (1.1).

If $w(x)$ is a positive continuously differentiable function on $[a,b]$ with $w(x) \leq c_1$ and $|w'(x)| \geq c_2 \geq 0$, then

$$(1.3) \quad \sum_{1 \leq i < j \leq N} m(I_i \cap I_j) \geq \frac{1}{2}\{\frac{L^2 h}{b-a} + \frac{L^2 h^2}{48} \frac{c_2^2}{c_1^3} \frac{b-a}{N^2} - \sum_{i=1}^{N} m(I_i)\}.$$

On the other hand, there will be infinitely many N for which

$$(1.4) \quad \sum_{1 \leq i < j \leq N} m(I_i \cap I_j) \leq \frac{1}{2}\{\frac{L^2 h}{b-a} + c_1(b-a) - \sum_{i=1}^{N} m(I_i)\}$$

when the intervals I_i are chosen appropriately.

In general, let m be a nonnegative measure on the set S. If S_i and S_j are subsets of S, we define $d = d(S_i, S_j)$, the distance between them, by

$$d = m(S_i \triangle S_j)$$

where $S_i \triangle S_j$ is the symmetric difference of S_i and S_j, i.e., the set of all points which belong to one or the other of them, but not both. It is easy to verify that d is a metric, provided that we identify sets whose symmetric difference has measure zero.

We comment that this notion of distance was used by John B. Kelly to establish his theory of hypermetric spaces ([1], [2] and [3]), and that equation (3.1) of [6] is closely related to Theorem 3.1 and equation (2.3) of Kelly's papers [1] and [2], respectively.

We shall begin by considering a finite set $S = \{x_1, \ldots, x_n\}$ of n elements. To each $x \in S$ we attach a positive weight $w(x)$ and write $w_i = w(x_i)$. For $T \subseteq S$ we define $m(T)$, the measure of T, by

$$m(T) = \sum_{x \in T} w(x)$$

Note that if $w(x) = 1$ for all $x \in S$ then

$$d(S_i, S_j) = m(S_i \triangle S_j) = |S_i \triangle S_j|,$$

the cardinality of the symmetric difference of S_i and S_j.

§2. The Distance Sum and the Discrepancy

Let $\sigma = (S_1, \ldots, S_N)$ be a sequence of subsets of the set S and let $r_i = |S_i|$. Define <u>distance sums</u>

$$(2.1) \quad \Sigma(\sigma) = \Sigma(N;\sigma) = \sum_{i<j} d(S_i, S_j)$$

and

$$(2.2) \quad \Sigma(N; r_1, \ldots, r_N) = \max \sum_{i<j} d(S_i, S_j)$$

where the maximum in (2.2) is taken over all sequences σ such that $|S_i| = r_i$ for $1 \leq i \leq N$. Define $f(x) = f(x;\sigma)$, the frequency of x in σ, to be the number of i such that $1 \leq i \leq N$ and $x \in S_i$. Write $f_i = f(x_i)$. We next let h be the harmonic mean of the weights w_i; i.e., $nh^{-1} = \sum_{i=1}^{n} w_i^{-1}$. We also define $|\sigma|$, the total cardinality of σ, by

$$|\sigma| = \sum_{i=1}^{N} r_i.$$

We now define $D(\sigma)$, the _discrepancy_ of σ, by

$$(2.3) \quad D(\sigma) = \sum_{j=1}^{n} \left(\frac{|\sigma|h}{nw_j} - f_j \right)^2 w_j .$$

Note that if $w_j = 1$ for all j, the typical element of this sum becomes

$$\left(\frac{|\sigma|}{n} - f_j \right)^2 ;$$

here $|\sigma|/n$ can be thought of as the expected number of times that x_j will occur in σ, while f_j is the actual number of times it occurs.

3. The Sum of Distances and Discrepancy is Constant

Given a sequence $\sigma = (S_1, \ldots, S_N)$ we define $m(\sigma)$, the measure of the sequence σ, by

$$m(\sigma) = \sum_{i=1}^{N} m(S_i) .$$

Theorem 3.1. For any sequence σ we have

$$(3.1) \quad \Sigma(\sigma) + D(\sigma) = Nm(\sigma) - \frac{|\sigma|^2 h}{n} .$$

Thus if we vary σ over all sequences having the same measure and the same total cardinality, the sum of the distances between elements of σ, and the discrepancy of σ, is constant.

Proof. Define a binary operation $*$ on the set $\{0,1\}$ as follows: $0*0 = 1*1 = 0$ and $0*1 = 1*0 = 1$. This is, of course, addition modulo 2. For vectors $y = (y_1, \ldots, y_n)$ and $z = (z_1, \ldots, z_n)$ whose components are zeros and ones, define

$$y*z = w_1(y_1*z_1) + \ldots + w_n(y_n*z_n) .$$

Define an $N \times n$ incidence matrix $A = (a_{ij})$ as follows:

$$(3.2) \quad a_{ij} = \begin{cases} 1 & x_j \in S_i \\ 0 & \text{otherwise.} \end{cases}$$

Let C_j denote the jth column of A, let $C_j(1)$ denote the number of ones in this column, and let $C_j(0)$ denote the number of zeros. Define $R_i, R_i(1)$, and $R_i(0)$ similarly. Clearly $R_i(1) = r_i$ and $C_j(1) = f_j$. Let $Q = \sum_{j=1}^{n} w_j C_j(1)^2$. Since $C_j(0) + C_j(1) = N$, we have

$$\Sigma(\sigma) = \sum_{i<k} R_i * R_k = \sum_{j=1}^{n} w_j C_j(0) C_j(1) = N \sum_{j=1}^{n} w_j C_j(1) - Q$$

$$= Nm(\sigma) - Q$$

Now by expanding (2.3) we find that

$$Q = D(\sigma) + \frac{|\sigma|^2 h}{n} ,$$

so the proof is complete.

Note. It follows immediately that $|\sigma|^2 h \leq n \cdot m(\sigma) N$; i.e.
$(\sum f_j)^2 \leq N (\sum w_j^{-1}) (\sum w_j f_j)$. This can also be seen directly since $|f_j| \leq N$ and

$$(\sum (f_j/w_j)^{1/2} (w_j f_j)^{1/2})^2 \leq (\sum f_j/w_j) (\sum w_j f_j).$$

There is also an integral analogue. Let $m*$ be a positive Borel measure on a subset S of E^n. Let $k > 0$ and let $w(x)$ be a bounded $m*$-measurable function defined on S such that $w(x) > k$ for $x \in S$. For $T \subseteq S$ let

$$(3.3) \quad m(T) = \int_T w(x) dm*(x)$$

and

$$(3.4) \quad h = m*(S) \{ \int_S \frac{dm*(x)}{w(x)} \}^{-1}.$$

For a sequence $\sigma = (S_1, \ldots, S_N)$ of subsets of S, define

$$(3.5) \quad L = L(\sigma) = \sum_{i=1}^{N} m*(S_i)$$

and let $f(x) = f(x;\sigma)$ be the number of i such that $1 \leq i \leq N$

and $x \in S_i$. The following results is an immediate consequence of Theorem 3.1.

<u>Theorem 3.2</u>. If $M*(S)$ is finite, then

$$(3.6) \quad \sum_{1 \leq i < j \leq N} m(S_i \, \Delta \, S_j) + \int_S \left(\frac{Lh}{m*(S)w(x)} - f(x) \right)^2 dm(x)$$

$$= N \sum_{i=1}^{N} m(S_i) - \frac{L^2 h}{m*(S)} \, .$$

We have made no effort to write down the most general such theorem; for example, one need not restrict S to be a subset of a Euclidean space. It is of greater importance to find some interesting special cases of (3.6). Let S be the interval $[a,b]$, let the S_i be intervals contained in S (of total length L) and let $w(x) \equiv 1$. Then $m = m*$ and

$$(3.7) \quad \sum_{1 \leq i < j \leq N} m(S_i \, \Delta \, S_j) + \int_a^b \left(\frac{L}{b-a} - f(t) \right)^2 dt = NL - \frac{L^2}{b-a} \, .$$

If $[a,b] = [0,1]$ and $S_i = [0,x_i]$, we obtain

$$(3.8) \quad \sum_{i<j} |x_i - x_j| + \int_0^1 (L - f(t))^2 \, dt = L(N - L);$$

here the integral measures to what extent the intervals $[x_i, x_j]$ fail to cover every point of $[0,1]$ the same number of times.

§4. <u>An Application</u>

Let S be the real interval $[a,b]$ where $0 < a < b$, and for $x \in S$ let $w(x)$ be a function which is positive and continuous on the closed interval $[a,b]$. Let $S_1, \ldots, S_N$ be subintervals of S of total length L.

<u>Problem</u>. How large can $\sum_{i<j} m(S_i \, \Delta \, S_j)$ be? In other words, how can you arrange these intervals so that they "avoid each other as much as possible (in the weighted sense)"?

What we need is a lower bound for D, the integral on the left of (3.6). In this section let $x_0 = a$, $x_n = b$, and $x_0 < x_1 < \cdots < x_{n-1} < x_n$ where $x_1, \ldots, x_{n-1}$ are the endpoints of intervals $S_1, \ldots, S_N$; thus $n \leq 2N + 1$. Clearly

$$(4.1) \quad D = \sum_{i=1}^{n} \int_{x_{i-1}}^{x_i} (a(x) - f_i)^2 \, dm(x)$$

where f_i is a constant and $a(x) = Lh/(b - a)w(x)$. Each term of the sum of (4.1) would only become smaller if each f_i were re-placed by

$$(4.2) \quad f_i' = \{\int_{x_{i-1}}^{x_i} a(x) dm(x)\} / \{\int_{x_{i-1}}^{x_i} w(x) dx\}.$$

Define

$$J_1(i) = \int_{x_{i-1}}^{x_i} \frac{dx}{w(x)} \, ,$$

$$J_2(i) = \int_{x_{i-1}}^{x_i} \lceil w(x) - \frac{x_i - x_{i-1}}{J_1(i)} \rceil^2 \frac{dx}{w(x)} \, ,$$

and

$$J_3(i) = \int_{x_{i-1}}^{x_i} w(x) dx.$$

Thus

$$D \geq \frac{L^2 h^2}{(b - a)^2} \sum_{i=1}^{n} \{\int_{x_{i-1}}^{x_i} \frac{dx}{w(x)} - \frac{(x_i - x_{i-1})^2}{J_3(i)}\}$$

$$= \frac{L^2 h^2}{(b - a)^2} \sum_{i=1}^{n} \frac{J_1(i) J_2(i)}{J_3(i)} \, .$$

Assume that $w(x)$ is a positive continuously differentiable func-tion on $[a,b]$ with $w(x) \leq C_1$ and $|w'(x)| \geq C_2 \geq 0$. Then for any u satisfying $x_{i-1} \leq u \leq x_i$, the mean value theorem yields

$$\int_{x_{i-1}}^{x_i} [w(x) - w(u)]^2 \frac{dx}{w(x)} = \int_{x_{i-1}}^{x_i} [w'(u(x))]^2 (x - u)^2 \frac{dx}{w(x)}$$

$$\geq \frac{C_2^2}{3C_1} [(x_i - u)^3 - (x_{i-1} - u)^3]$$

$$\geq \frac{c_2^2}{12c_1} (x_i - x_{i-1})^3 .$$

Also

$$\sum_{i=1}^{n} (x_i - x_{i-1})^3 \geq (n - 1)(\frac{b - a}{n - 1})^3 = \frac{(b - a)^3}{(n - 1)^2} .$$

We can now obtain a lower bound on J_2, since $x_{-1} \leq u \leq x_i$ holds for some u satisfying $w(u) = (x_i - x_{i-1})/J_1(i)$. Thus

$$D \geq \frac{L^2 h^2}{(b - a)^2} \sum_{i=1}^{n} \frac{c_1^{-1}(x_i - x_{i-1})}{c_1(x_i - x_{i-1})} \frac{c_2^2}{12c_1} (x_i - x_{i-1})^3$$

$$\geq \frac{L^2 h^2 (b - a) c_2^2}{12 c_1^3 (n - 1)^2} \geq \frac{L^2 h^2}{48} \frac{c_2^2}{c_1^3} \frac{b - a}{N^2} .$$

We have proved the following.

<u>Theorem</u>. Let $w(x)$ be a positive continuously differentiable function on $[a,b]$ with $w(x) \leq C_1$ and $|w'(x)| \geq C_2 \geq 0$. Then

$$(4.3) \quad \sum_{1 \leq i < j \leq N} m(S_i \triangle S_j) \leq N \sum_{i=1}^{N} m(S_i) - \frac{L^2 h}{b - a} - \frac{L^2 h^2 c_2^2}{48 c_1^3} \frac{b - a}{N^2} .$$

Next, from the identity

$$(4.4) \quad 2m(S_i \cap S_j) = m(S_i) + m(S_j) - m(S_i \triangle S_j)$$

we easily derive

$$(4.5) \quad \sum_{i-j} m(S_i \triangle S_j) = (N - 1) \sum_{i=1}^{N} m(S_i) - 2 \sum_{i<j} m(S_i \cap S_j);$$

inequality (1.3) now follows from (4.3) and (4.5). Inequality (1.2) follows similarly: use the trivial inequality $D \geq 0$.

We shall now show that $D \leq C_1(b - a)$ for infinitely many N when the intervals I_j are chosen appropriately. Let $a = x_0 < x_1 < \ldots < x_{n-1} < x_n = b$ be a partition of $[a,b]$ such that the difference between the maximum and minimum value of the $a(x)$ of (4.1) is less than 1 on any interval $[x_{i-1}, x_i]$. Let f_i be an integer such that $|f_i - a(x)| \leq 1$ for $x_{i-1} \leq x \leq x_i$. Let the intervals I_j consist of f_1 copies of $[x_0,x_1]$, f_2 copies of $\lceil x_1,x_2 \rceil, \ldots,$ and f_n copies of $\lceil x_{n-1},x_n \rceil$. Then by (4.1) we have

$$D \leq \sum_{i=1}^{n} \int_{x_{i-1}}^{x_i} w(x)dx \leq C_1(b - a),$$

and (1.4) is proved.

§5. A "Symmetric" Generalization

The theorem of Section 3 is not completely satisfactory, since it considers the set S as weighted but the subsets S_i as unweighted. We shall now consider sequences of weighted subsets of S, or, what is the same thing, sequences of weights defined on S.

We say a function m_i with domain S is a weight defined on S if for every $x \in S$ the function value $m_i(x)$ is a nonnegative real number. We define $d = d(m_i,m_j)$, the distance between the weights m_i and m_j, by

$$(5.1) \quad d(m_i,m_j) = \sum_{x \in S} |m_i(x) - m_j(x)| w(x)$$

where $w(x)$ is the original weight defined on S. Clearly d is a metric. If for $S_i \subseteq S$, we have $m_i(x) = 1$ for $x \in S_i$ and $m_i(x) = 0$ otherwise, then $d(m_i,m_j) = m(S_i \triangle S_j)$ and we are back in the previous situation. For a sequence $\sigma = (m_1,\ldots,m_N)$ of N weights, we define the distance sum $\Sigma(\sigma)$ by

$$(5.2) \quad \Sigma(\sigma) = \sum_{i<j} d(m_i,m_j).$$

Next, let

$$(5.3) \quad f(x) = \sum_{i=1}^{N} m_j(x);$$

again we can call $f(x)$ the frequency of x in σ. We also need to measure the frequency with which the nonnegative real number t is expected to occur as a weight. Let $\emptyset_x(t)$ denote the number of i with $1 \leq i \leq N$ such that $0 \leq t \leq m_i(x)$. We may call f and $\emptyset_x$ the element frequency and weight frequency respectively. Let

$$L = L(\sigma) = \sum_{x \in S} f(x)$$

and

$$m(\sigma) = \sum_{x \in S} w(x) f(x).$$

We call L and $M(\sigma)$ the length (or total cardinality) and measure of the sequence σ, respectively.

Now define

$$(5.4) \quad D_E(\sigma) = \sum_{x \in S} (\frac{Lh}{nw(x)}) - f(x))^2 w(x)$$

where $nh^{-1} = \sum_{x \in S} w(x)^{-1}$; we call $D_E(\sigma)$ the <u>element discrepancy</u> of σ. Finally, define

$$(5.5) \quad D_W(\sigma;b) = \sum_{x \in S} w(x) \int_O^b \frac{f(x)}{b} - \emptyset_x(t))^2 dt$$

where b is an upper bound for $\max m_i(x)$, the maximum being taken over all $x \in S$ and all i such that $1 \leq i \leq N$. We call $D_W(\sigma;b)$ the <u>weight discrepancy</u> of σ for $[0,b]$.

Theorem. If S is a finite set of n elements, then

$$(5.6) \quad \Sigma(\sigma) + \frac{1}{b} D_E(\sigma) + D_W(\sigma;b) = Nm(\sigma) - \frac{L^2 h}{nb} .$$

<u>Proof</u>. In (3.7) let $S_i = S_i(x) = [0, m_i(x)]$ and let $a = 0$. Also, replace the f in (3.7) by $\emptyset_x$. It now follows from (3.7) and (5.3) that

$$(5.7) \quad \Sigma(\sigma) = \sum_{x \in S} w(x) \sum_{i < k} |m_i(x) - m_k(x)|$$

$$= \sum_{x \in S} w(x) \{Nf(x) - \frac{f^2(x)}{b} - \int_O^b (\frac{f(x)}{b} - \emptyset_x(t))^2 dt\}$$

$$= Nm(\sigma) - D_W(\sigma;b) - \frac{1}{b} Q$$

where

$$Q = \sum_{x \epsilon S} w(x) f^2(w).$$

By expanding (5.4) we find that

$$Q = D_E(\sigma) + \frac{L^2 h}{n}$$

and this completes the proof.

We can slightly improve the appearance of (5.6) by only considering weights between 0 and 1, so that we can take b = 1.

§6. Remark

Problems concerning irregularities of distribution arise in several areas of mathematics. K.F. Roth's paper [4] is especially noteworthy; his work has been greatly extended recently by W.M. Schmidt. For an excellent general survey of this area, together with proofs of many of its most outstanding results, see [5]. Paper [6] relies heavily on Schmidt's results for irregularities of distribution on spheres to estimate the distance sums involved.

REFERENCES

[1] J.B. Kelly, Metric inequalities and symmetric differences. In
     "Inequalities - II" (O. Shisha, ed.), pp. 193-212. Academic
     Press, New York, 1970.

[2] J.B. Kelly, Combinatorial inequalities. In "Combinatorial
     Structures and Their Applications" (R. Guy, H. Hanani, N.
     Sauer, and J. Schonheim, eds.), pp. 201-207. Gordon and
     Breach, New York, 1970.

[3] J.B. Kelly, Hypermetric spaces and metric transforms. In
     "Inequalities - III" (O. Shisha, ed.), pp. 149-158. Academic
     Press, New York, 1972.

[4] K.F. Roth, On irregularities of distribution, Mathematika 7
     (1954), 73-79.

[5] W.M. Schmidt, Lectures on Irregularities of Distribution,
     University of Colorado, Boulder, 1973.

[6] K.B. Stolarsky, Sums of distances between points on a sphere
     II, Proc. Amer. Math. Soc. 41 (1973), 575-582.

METRIC EMBEDDING TECHNIQUES APPLIED TO GEOMETRIC INEQUALITIES

Ralph Alexander
University of Illinois

Dedicated to Professor Leonard M. Blumenthal

ABSTRACT

A method for "metric addition" of simplices in Euclidean space is developed and applied to certain geometric inequalities. In the case of two simplicies with vertices $p_0, \ldots, p_{n-1}$ and $p_0', \ldots p_{n-1}'$ respectively, a new simplex $p_0'', \ldots, p_{n-1}''$ is made such that $|p_i'' - p_j''|^2 = \frac{1}{2}[|p_i - p_j|^2 + |p_i' - p_j'|^2]$ for all i, j. The Cauchy-Schwarz inequality then implies that $|p_i'' - p_j''| \geq \frac{1}{2}[|p_i - p_j| + |p_i' - p_j'|]$.

This construction, together with a lemma of G.T. Sallee, allows one to prove discrete analogues of various chord and arc-length inequalities. For example, if $p_0, \ldots, p_{n-1}$ are points in Euclidean space such that $|p_i - p_{i+1}| = 1$ for each $i \pmod n$, then the sum $\Sigma_{i,j} |p_i - p_j|$ is uniquely maximal when the points are the distinct vertices of a regular n-gon.

A metric embedding theorem of I.J. Schoenberg allows a continuous version of the method. If $r(t): 0 \leq t < \pi$, is a closed curve in ℓ_2 which is parametrized by arclength modulo π, it is shown that there is another such curve $r'(t)$ such that $|r'(t_1) - r'(t_2)|^2 = \frac{1}{\pi} \int_0^\pi |r(t_1 + t) - r(t_2 + t)|^2 dt$ for all t_1, t_2. The fact that the curve $r'(t)$ contains the circle group in its group of isometries is of crucial importance.

If s, t vary over the real plane modulo π, then the method easily demonstrates that the function $I(r) = \iint |r(t) - r(s)| ds\, dt$ is maximal when $r(t)$ is the boundary of a circle.

§1. Introduction

Recently, there have been a number of articles written on questions related to a problem of H. Herda. The note by Herda [3] contains a long list of references to the many and varied solutions which have been offered.

In the original problem we suppose that $C:r(t)$, $0 \leq t \leq \pi$, is a closed curve of length π lying in a Euclidean space or, perhaps, in a Hilbert space. We assume that C is parametrized by arclength modulo π. We are asked to show that the minimum of the $|r(t) - r(t + \pi/2)|$ is at most 1, with equality holding only if C is a circle.

We mention a second problem: Show that the integral $\iint |r(s) - r(t)| ds\, dt$ is uniquely maximal when C is a circle. Here (s,t) varies over the real plane modulo π.

In this article we would like to point out a very effective approach to many such problems. The method combines a lemma of G.T. Sallee with a criterion for metric embedding into the classical Hilbert space ℓ_2 due to I.J. Schoenberg. The book of L.M. Blumenthal [1], especially chapter 4, is the standard reference on metric embedding into Euclidean and Hilbert space.

Since the various approaches to the Herda problem which we have seen involve continuous (or differentiable) techniques, we feel that is is appropriate to formulate and solve by means of distance geometry discrete versions of the two problems above. This will expose the method while pointing out inequalities not easily obtained by other methods. However, in Lemma 2 we show that the approach is also effective for continuous problems.

§2. Two Discrete Problems

A collection of $n(n \geq 2)$ points $r_0, r_1, \ldots, r_{n-1}$ in a Euclidean space, indexed by residue classes modulo n, such that

$|r_{i+1} - r_i| = 1$ will be called an $\underline{n\text{-chain}}$. The n-chain will be called $\underline{\text{convex}}$ if the closed polygonal line, whose successive vertices are $r_0, \ldots, r_{n-1}$, bounds a plane convex body. (Here we assume a winding number 1, of course.)

$\underline{\text{Problem 1}}$. Let $r_0, \ldots, r_{n-1}$ be an n-chain with n even. Show that the minimum of the distances $|r_i - r_{i+n/2}|$ is at most equal to the diameter of the regular n-gon of perimeter n, with equality holding only if the r_i are vertices of the regular n-gon of perimeter n.

$\underline{\text{Problem 2}}$. Let $r_0, \ldots, r_{n-1}$ be an n-chain. Show that the sum $\sum_{i,j} |r_i - r_j|$ is uniquely maximal when the r_i are the vertices of a regular n-gon of perimeter n.

It should be noted that if only the requirement $\sum_i |r_{i+1} - r_i| = n$ is imposed on $r_0, \ldots, r_{n-1}$, then the conclusions of these problems become false as simple examples with $n = 4$ show.

The following lemma of G.T. Sallee [4] is essential to the remainder of our article. Although his lemma is true for arbitrary closed polygonal curves, we give a statement in terms of n-chains.

$\underline{\text{Lemma 1}}$. (Sallee) Let $r_0, \ldots, r_{n-1}$ be an n-chain which is not convex. Then there exists a convex n-chain $r_0', \ldots, r_{n-1}'$ such that $|r_i' - r_j'| > |r_i - r_j|$ unless i and j differ by one modulo n.

The following theorem occurs in essence in the paper [5] of I.J. Schoenberg.

$\underline{\text{Theorem 1}}$. (Schoenberg) Let $(\mathcal{m}, d)$ be a separable semimetric space. Then $(\mathcal{m}, d)$ can be isometrically embedded in the sequence space ℓ_2 if and only if for any finite collection of points $p_1, \ldots, p_n$ in $\mathcal{m}$, the quadratic form $\sum_{i,j} d(p_i, p_j)^2 x_i x_j$ is negative semidefinite on the hyperplane $\sum_i x_i = 0$.

$\underline{\text{Corollary 1}}$. Let $r_0, \ldots, r_{n-1}$ be distinct points which form an n-chain. Then there exists an n-chain $r_0', \ldots, r_{n-1}'$ such that

(1) $\quad |r_i' - r_j'| = [\frac{1}{n} \sum_{k=0}^{n-1} |r_{i+k} - r_{j+k}|^2]^{1/2}$

(2) $\quad |r_i' - r_j'| \geq \frac{1}{n} \sum_{k=0}^{n-1} |r_{i+k} - r_{j+k}|$

with equality holding only if $|r_i - r_j| = |r_{i+k} - r_{j+k}|$ for each k.

Proof. The quadratic form $\sum_{i,j} |r_i - r_j|^2 x_i x_j$ is negative semidefinite on the hyperplane $\sum_i x_i = 0$. The qaudratic form whose coefficients are given by $|r_i' - r_j'|^2$ has this property also, and defines a semimetric on an n-element set by means of equation (1). Theorem 1 assures us that there exist points $r_0', r_1', \ldots, r_{n-1}'$ in ℓ^2, hence in E^{n-1}, such that the Euclidean metric agrees with the semimetric given by equation (1). Since $|r_{i+1}' - r_i'| = 1$ for each i, the points form an n-chain.

The inequality (2) follows at once upon application of the Cauchy-Schwarz inequality to equation (1). This completes the proof.

We should point out that equation (1) is only one of many ways in which root mean square averaging of distances might be done. Equation (6) below gives another.

The RMS averaging process allows us to introduce symmetry in such a way that many interesting distance relationships are retained. We note that the distinct points $r_0', \ldots, r_{n-1}'$ obtained in Corollary 1 have a symmetry group which contains the n-element cyclic group.

Let us apply these results to our problems, taking the second problem first. By a standard compactness argument an n-chain $r_0, \ldots, r_{n-1}$ will exist in E^{n-1} such that the sum $\sum_{i,j} |r_i - r_j|$ is maximal. Lemma 1 assures us that this n-chain is convex. We apply the RMS average (1) to obtain the points $r_0', \ldots, r_{n-1}'$ in

E^{n-1}. The inequality (2), summed over all i,j, assures us that

(3) $\quad \Sigma_{i,j} |r'_i - r'_j| \geq \Sigma_{i,j} |r_i - r_j|$,

and equality can hold only if for every i,j,k,

(4) $\quad |r_{i+k} - r_{j+k}| = |r_i - r_j|$.

However, equation (4) immediately implies that $|r_i - r_j|$ equals $|r'_i - r'_j|$ for all i,j. A convex n-chain which contains the n-element cyclic group in its symmetry group clearly bounds a regular n-gon of perimeter n.

The argument for the discrete Herda problem is only slightly more involved. Again, let us suppose that $r_0, \ldots, r_{n-1}$ are chosen so that the minimum of the distances $|r_i - r_{i+n/2}|$ is maximal. As before this n-chain must be convex. We apply equation (1) to obtain the n-chain $r'_0, \ldots, r'_{n-1}$ in E^{n-1}. This n-chain must be convex, or we could apply Lemma 1 to obtain a contradiction to the assumed extremal property of $r_0, \ldots, r_{n-1}$. Inequality (2) allows us to conclude for each i,

(5) $\quad |r'_i - r'_{i+n/2}| \geq \min_j |r_j - r_{j+n/2}|$.

Because of the assumed extremality, the inequality (5) must in fact be equality. As before, the symmetry of the convex n-chain $r'_0, \ldots, r'_{n-1}$ forces it to be bound a regular n-gon of perimeter n. We have established the desired inequality. Also, all distances $|r_i - r_{i+n/2}|$ must be equal in an extremal n-chain.

Now let us assume that $n \geq 4$. To establish the uniqueness of the extremal configuration, we again apply Theorem 1 to assert that there are points $r''_0, \ldots, r''_{n-1}$ in E^{n-1} such that

(6) $\quad |r''_i - r''_j| = [\frac{1}{2}|r_i - r_j|^2 + \frac{1}{2}|r_{i+n/2} - r_{j+n/2}|^2]^{1/2}$.

As before the n-chain $r_0'', \ldots, r_{n-1}''$ must be convex. Since its dia-gonals are equal, the parallelogram $r_i'', r_{i+1}'', r_{i+n/2}'', r_{i+n/2+1}''$ must in fact be a rectangle. Let us say that equation (6) (together with elementary trigonometry) forces the points $r_i, r_{i+1}, r_{i+n/2}, r_{i+n/2+1}$ to be the vertices of a rectangle. Since the diagonals of these rectangles all contain a common midpoint, it is clear that $r_0, \ldots, r_{n-1}$ are the vertices of a regular n-gon of perimeter n.

We should note that the integer n/2 can be replaced by any integer m not congruent to 1 modulo n, and it would follow that the minimum of the distances $|r_i - r_{i+m}|$ is maximal when the n-chain bounds a regular n-gon. However, some further discussion would be required to establish the uniqueness of the configuration.

§3. The Continuous Version of the Method

A closed curve $\mathcal{C}$ of length π lying in ℓ_2 without self-intersection will be termed proper. The curve will be parameteriz-ed by arclength modulo π.

Lemma 2. Let $\mathcal{C}$ be a proper curve. Then a proper curve $\mathcal{C}'$ exists such that

$$(7) \quad |r'(t_1) - r'(t_2)| = [\frac{1}{\pi} \int_0^\pi |r(t_1 + t) - r(t_2 + t)|^2 \, dt]^{1/2},$$

and

$$(8) \quad |r'(t_1) - r'(t_2)| \geq \frac{1}{\pi} \int_0^\pi |r(t_1 + t) - r(t_2 + t)| \, dt$$

with equality holding in (8) only if the integrand is constant.

Proof. Equation (7) defines a separable semimetric on a set indexed by the reals modulo π. Also, it is clear that if $t_1, \ldots, t_2$ are numbers modulo π, the quadratic form $\sum_{i,j} |r'(t_i) - r'(t_j)|^2 x_i x_j$ is negative semidefinite on the hyper-plane $\sum_i x_i = 0$ because the integral average preserves this pro-perty. Thus there is a closed curve $\mathcal{C}'$ in ℓ_2 such that equation

(7) gives the distance between any two points on c'. The inequality (8) follows at once from the Cauchy-Schwarz inequality.

If t_0 is a real number, let $t_k = t_0 + \frac{\pi k}{n}$ for $k = 0, \ldots,$ n-1. Since for any t_0, $|r(t_{i+1}) - r(t_i)| \leq \frac{\pi}{n}$, we can choose $n(\varepsilon)$ such that for any t_0

$$(9) \quad \pi - \varepsilon \leq \sum_{i=0}^{n-1} |r(t_{i+1}) - r(t_i)| \leq \pi.$$

Let us define in $s_i = t_i + t$ and consider $\frac{1}{\pi} \int_0^\pi \sum_{i=0}^{n-1} |r(s_{i+1}) - r(s_i)| dt$. The left inequality in (9) together with inequality (8) imply that $\sum_{i=0}^{n-1} |r'(t_{i+1}) - r'(t_i)| \geq \pi - \varepsilon$. On the other hand since $|r(s_{i+1}) - r(s_i)| \leq \frac{\pi}{n}$ for all t, t_0, $|r'(t_{i+1}) - r'(t_i)| \leq \frac{\pi}{n}$ by equation (7). Letting n tend to infinity shows that c' in fact has length π. A modification of the above argument, applied to arcs of c', shows that $r'(t)$ is itself a parameterization of c' by arclength modulo π

The equation (7) clearly shows that the curve c' contains a copy of the circle group in its group of isometries.

Although the details have not been written down, we claim that if c is a closed rectifiable curve in ℓ_2 which does not bound a plane convex body, then there exists a curve c^* which is a strictly stretched (in the sense of Sallee [4]) version of c. If one is not interested in the uniqueness of extremal configurations, but only in inequalities, then a stretched version of c will suffice.

The method works along the same pattern as in the discrete case. For example, one easily shows that if $0 \leq c \leq \pi/2$ then $\min_r |r(t) - r(t + c)| \geq \sin c$. The question of this inequality was raised by H.H. Johnson according to Herda [3].

§4. Further Problems

One could pose many problems related to RMS averaging of distances. We mention one: Suppose two simplices have vertices

$p_1, \ldots, p_n$ and $p_1', \ldots, p_n'$. Form the simplex $p_1'', \ldots, p_n''$ such that $|p_i'' - p_j''|^2 = \frac{1}{2}[|p_i - p_j|^2 + |p_i' - p_j'|^2]$. We conjecture that $v''^2 \geq \frac{1}{2}[v^2 + v'^2]$, where V, V', V'' are the respective volumes.

The following problem seems difficult: Suppose C is a closed rectifiable curve of length π. If x, y are on C, define $\delta_x = \max_y |x - y|$. Is it true that $\min \delta_x \leq 1$, with equality only if C bounds a plane body of constant width? Chakerian [2] proves an inequality which is somewhat similar, but his method doesn't seem to work here. The case where C is a square seems to provide a counterexample to many possible approaches.

In closing, I thank Dr. Larry Lipskie, whose elegant solution of Herda's problem, aroused my interest in these matters, and I also thank Professor G.T. Sallee for kindly sending me preprints of his work.

ADDENDUM

Recently, we have noted a simple constructive proof of Corollary 1. Let $r_0, r_1, \ldots, r_{n-1}$ lie in E^d and consider the following set of points in E^{nd}:

$$r_0' = \frac{1}{\sqrt{n}} (r_0, r_1, \ldots, r_{n-1}),$$

$$r_1' = \frac{1}{\sqrt{n}} (r_1, r_2, \ldots, r_{n-1}, r_0), \ldots$$

$$r_{n-1}' = \frac{1}{\sqrt{n}} (r_{n-1}, r_0, \ldots, r_{n-2}).$$

The Pythagorean theorem assures us that equation (1) is satisfied by the points $r_0', \ldots, r_{n-1}'$. This observation will certainly make any future study of RMS averaging of distances much easier, at least for the discrete case. Schoenberg's embedding theorem still seems to be the only reasonable way to justify the continuous version of RMS averaging of distances.

REFERENCES

[1] L.M. Blumenthal, <u>Theory and applications of distance geometry</u>, Clarendon Press, Oxford, 1953.

[2] G.D. Chakerian, <u>A characterization of curves of constant width</u>, Amer. Math. Monthly 81 (1974), 153-155.

[3] H. Herda, <u>A characterization of circles and other closed curves</u>, Amer. Math. Monthly 81 (1974), 146-149.

[4] G.T. Sallee, <u>Stretching chords of space curves</u>, Geometriae Dedicata 2 (1974), 311-317.

[5] I.J. Schoenberg, Remarks to Maurice Fréchet's article "<u>Sur la definition axiomatique d'une classe d'espaces vectoriels distancies applicables vectoriellment sur l'espace de Hilbert</u>," Ann. of Math. 36 (1935), 724-732.

ANGLES IN METRIC SPACES

Joseph E. Valentine
Utah State University

ABSTRACT

If a,b,c are points of a metric space, then it follows from the triangle inequality that

$$-1 \leq (ab^2 + ac^2 - bc^2)/2ab \cdot ac \leq 1,$$

where juxtaposition denotes the distance between two points. Thus calling the symbol bac an angle with vertex a, its value is given by the formula

(1) $bac = \operatorname{Arc\ cos} [(ab^2 + ac^2 - bc^2)/2ab \cdot ac].$

Wilson introduced this concept of angle in general metric spaces. Of course there may be no significant relationship between bac and dae even when d and e are points between a and b and a and c, respectively. Wilson overcame this objection by defining an angle (ρ,σ) of two rays (congruent images of half-lines) ρ,σ with common initial point a by

(2) $(\rho,\sigma) = \lim bac$

as b,c tend to a on the ray ρ and σ, respectively, provided this limit exists. Another way of making angles well defined is to insist the euclidean law of cosines (1) be homogeneous in the space; that is, require bac = bae whenever e is any point on a ray joining a and c and also require $bac = \pi - baf$ whenever f is any point on a line containing the ray joining a and c but such that a is between c and f.

We show that a complete, convex, externally convex metric space is an inner-product space if and only if the cosine law is homogeneous for some particular angle different from 0 and π. Whether or not the requirement that particular angles and their supplements between a ray and a line be defined as in (2) characterizes inner-product space among the class of complete, convex, externally convex metric spaces remains unknown. The convexly metrized tripod shows the supplement condition or some other condition must be imposed. Two closely related questions which arise in this investigation are stated.

ANGLES IN METRIC SPACES

Joseph E. Valentine
Utah State University

Blumenthal [3] said the pythagorean theorem is valid in a metric space M provided for any line L of M and for any point p of M, p not on L, $pp_o^2 + p_o x^2 = px^2$ for all x on L, where p_o is a foot of p on L. (For a detailed study of metric concepts and notation used in this paper, the reader is referred to [2].) He proved a complete, convex, externally convex metric space M is an inner-product space if and only if the pythagorean theorem is valid in M. He accomplished this by showing the validity of the pythagorean theorem in M implies M has the euclidean weak four-point property. This completes the characterization, for Blumenthal and previously shown the euclidean weak four-point property characterizes euclidean space among the class of complete, convex, externally convex metric spaces.

James [4] said vectors x,y in a real normed linear space are orthogonal in the sense of Pythagoras provided $\|x\|^2 + \|y\|^2 = \|x - y\|^2$. He showed a real normed linear space is an inner-product space if and only if pythagorean orthogonality is homogeneous; that is,

$$\|x\|^2 + \|y\|^2 = \|x - y\|^2$$

implies

$$\|x\|^2 + \|\beta y\|^2 = \|x - \beta y\|^2$$

for all real β.

Both of these ideas can be stated in terms of the euclidean law of cosines. Thus the pythagorean theorem is valid in M provided for any line L of M and any point p of M, p not on L, the cosine of the angle with vertex p_o and sides the segments

joining p_0 and p and p_0 and x, respectively, (as given by
the euclidean law of cosines) is zero, where p_0 is a foot of p
on L and x is any point on L. In case M is a real normed
linear space, x,y are orthogonal provided the cosine of the angle
with vertex θ and sides the segments joining θ and x and θ
and y, respectively, is zero. Moreover, homogeneity of phthagor-
ean orthogonality is equivalent to the cosine of the angle with
vertex θ and sides the algebraic segments joining θ and x and
θ and βy being zero for all real β.

Martin and Valentine [5] have obtained reasonably definitive
results concerning angles defined by the euclidean law of cosines
in real normed linear spaces. They have obtained some results in
the setting of complete, convex, externally convex metric spaces.
In the following we briefly outline the latter results and mention
some questions which arise naturally in this context. We need the
following two definitions.

Definition. If a,b,c are collinear points the signum of the
ordered triple (a,b,c) is defined by

$$\text{sgn}(a,b,c) = \begin{cases} 1 & \text{if } c \text{ is between } a \text{ and } b \text{ or } a \text{ is} \\ & \quad \text{between } b \text{ and } c \\ \\ -1 & \text{if } b \text{ is between } a \text{ and } c. \end{cases}$$

Definition 2. Let k be an arbitrary but fixed number,
$-1 \leq k \leq 1$. The euclidean law of cosines is k-signum homogeneous
in M provided if p,q,r are distinct elements of M with

$$[pq^2 + pr^2 - qr^2]/2pq \cdot pr = k,$$

then for each q' on any ray with initial point p and passing
through q and for each r' on any line passing through p and
r,

$$pq'^2 + pr'^2 - q'r'^2 = 2pq' \cdot pr' \cdot k \ \text{sgn}(r', p, r).$$

The following lemma is of a technical nature and its proof depends on elementary limit type arugments.

Lemma. If L is a line of M, q a point of M, q not on L, and if $-1 < k < 1$, then L contains points p, r such that

$$[pq^2 + pr^2 - qr^2]/2pq \cdot pr = k.$$

Theorem. A complete, convex, externally convex metric space M is an inner-product space if and only if the euclidean law of cosines is k-signum homogeneous in M for some k, $-1 < k < 1$.

Proof. By the above mentioned result of Blumenthal, it suffices to show each quadruple of points of M that contains a linear triple is congruent to a quadruple of points of the euclidean plane E_2.

Let p, q, r, s be points of M with p, r, s linear. It is known that M contains a metric line L which contains p, r, and s. If q is on that line, then the quadruple p, q, r, s is congruent to a quadruple of E_2. If q is not on L, use the lemma to find points t, u on L such that

$$[qt^2 + tu^2 - qu^2]/2qt \cdot tu = k.$$

Let q', t', u' denote points in E_2 which are congruent to q, t, u. Use of the k-signum homogeneity of the cosine law easily shows the quadruple p, q, r, s is congruent to the quadruple p', q', r', s' where p', r', s' are points in E_2 on the line joining t', u' for which the quintuple s', r', p', t', u' is congruent to the quintuple s, r, p, t, u.

Denoting the angle

$$\text{Arc cos } [(ab^2 + ac^2 - bc^2)/2ab \cdot ac]$$

by bac, where a,b,c are points in a metric space, Wilson [7] defined the angle (ρ,σ) of two metric rays, ρ,σ with common initial point a and passing through b and c, respectively, by

$$(\rho,\sigma) = \lim bac$$

as b,c tend to a on the rays ρ and σ, respectively, provided this limit exists.

It is known that the existence of this type angle for each pair of rays with common initial point in a real normed linear space implies the space is an inner-product space; for example, see [6].

In [5] it is shown that if there is some k, $-1 < k < 1$, such that in a real normed linear space A, whenever x,y are vectors of A for which

$$[\|x\|^2 + \|y\|^2 - \|x - y\|^2]/2\|x\| \cdot \|y\| = k,$$

then

$$\lim_{\tau,\sigma \to 0} [\|\tau x\|^2 + \|\sigma y\|^2 - \|\tau x - \sigma y\|^2)/2\|\tau x\| \cdot \|\sigma y\| - k\, \mathrm{sgn}(\tau\sigma)] = 0$$

then A is an inner-product space.

In a metric space M the above hypothesis becomes, there is some k, $-1 < k < 1$, such that whenever p,q,r are points of M for which

$$[pq^2 + pr^2 - qr^2]/2pq \cdot pr = k,$$

then

$$\lim_{q',r' \to p} [(pq'^2 + pr'^2 - q'r'^2)/(2pq' \cdot pr') - k\, \mathrm{sgn}(qpq')] = 0,$$

r' on a ray with initial point p and passing through r and q' on a line through p and q.

Whether or not this latter property characterizes inner-product spaces among the class of complete, convex, externally convex metric spaces remains an open question.

It can be shown that if the above hypothesis is assumed and one chooses sequences $\{q_n\}$ and $\{r_n\}$ on the segments joing p and q and p and r, respectively, such that $pq = 2^n pq_n$, and $pr = 2^n pr_n$, then $qr = 2^n q_n r_n$ for all but a finite number of n.

Andalafte and Blumenthal [1] said a metric space satisfies the Young postulate provided for each triple of points p,q,r of the space, if q',r' are the respective midpoints of p and q and p and r, respectively, then $qr = 2q'r'$.

They showed a complete, convex, externally convex metric space with the two-triple property is a Banach space if and only if it satisfies the Young postulate.

The above observation leads to the following two questions which seem to be of interest besides leading to a solution to the above question.

1. Does the local Young postulate characterize Banach spaces among the class of complete, convex, externally convex metric spaces with the two-triple property?

2. Let M be a complete, convex, externally convex metric space with the two-triple property. Suppose there is some k, $-1 < k < 1$ such that whenever p,q,r are elements of M with $[pq^2 + pr^2 - qr^2]/2pq \cdot pr = k$ and if q',r' are the respective midpoints of p and q and p and r, then $qr = q2'r'$. Is M a Banach space?

REFERENCES

[1]   E.Z. Andalafte and L.M. Blumenthal, Metric characterizations
      of Banach and euclidean spaces, Fund. Math. 55 (1964), 24-55.

[2]   L.M. Blumenthal, Theory and applications of distance geometry,
      Clarendon Press, Oxford, 1953.

[3]   _____, Distance geometry notes, Bull. Amer. Math.
      Soc. 50 (1944), 235-241.

[4]   R.C. James, Orthogonality in normed linear spaces, Duke. Math.
      J. 12 (1945), 291-302.

[5]   C. Martin and J.E. Valentine, Angles in metric and normed
      linear spaces (submitted for publication).

[6]   J.E. Valentine and S.G. Wayment, Wilson angles in linear
      normed spaces, Pac. J. Math. 36 (1971), 239-243.

[7]   W.A. Wilson, On certain types of continuous transformations of
      metric spaces, Amer. J. Math. 57 (1935), 62-68.

GEOMETRIC FIXED POINT THEORY AND INWARDNESS CONDITIONS

James Caristi and William A. Kirk
The University of Iowa

ABSTRACT

Let X be a Banach space, $D \subset X$, $T:D \to X$. If each point $x \in D$ is mapped into the closure of the set determined by those rays emanating from x which pass through other points of D, then T is said to be weakly inward on D. B. Halpern introduced such inwardness assumptions in 1965 in generalizing the Schauder-Tychonov theorem and subsequently many similar assumptions have arisen in the study of fixed point theory of contraction and non-expansive mappings. The main result of this paper is the following theorem: Let (M,d) be a complete metric space and $f:M \to M$. If there exists a lower semicontinuous function $\emptyset$ mapping M into the nonnegative real numbers such that for each $x \in M: d(x,f(x)) \leq \emptyset(x) - \emptyset(f(x))$, then f has a fixed point in M. It is shown how this theorem yields essentially all the known inwardness results of geometric fixed point theory in Banach spaces including results heretofore derived only from existence theory of ordinary differential equations. The main result typically applies to discontinuous mappings f; its proof describes a transfinite iterative procedure which always converges to a fixed point; and in addition it includes Banach's contraction mapping theorem as a special case.

GEOMETRIC FIXED POINT THEORY AND INWARDNESS CONDITIONS

James Caristi and William A. Kirk [1]
The University of Iowa

1. Introduction

Let X be a Banach space, $D \subset X$, $T:D \to X$. An inwardness condition on T is one which asserts that points $x \in D$ are mapped into the closure of the set determined by those rays emanating from x which pass through other points of D. Rather mild conditions of this type are often sufficient to assure existence of fixed points for mappings T which are not self-mappings on their domains. Precise inwardness assumptions were introduced by B. Halpern in 1965 [4] in his work generalizing the Schauder-Tychonov theorem (also see [5]). Since that time a number of inwardness results have been established for many of the important classes of mappings in Banach spaces, including in particular the contraction and nonexpansive mappings. These results have in fact been established in two quite different ways. In addition to a rather direct approach, for example as found in papers of S. Reich [8,9,10], even sharper fixed point theorems have been derived from certain existence theorems of ordinary differential equations for mappings satisfying an implicit inwardness condition (condition (2) below). This latter technique is found in work of R.H. Martin [6] and G. Vidossich [11].

In this report we discuss an improved version of a new metric space fixed point theorem of Caristi [3] and illustrate how it implies essentially all the known inwardness results for contraction and nonexpansive mappings. In addition to being an explicit generalization of Banach's contraction mapping theorem, the main result describes an iterative procedure (transfinite) which always

_____

[1]Research supported in part by the Louis Block Fund, The University of Chicago.

converges to a fixed point. This theorem applies to mappings which *typically* are not even continuous; in the special case where the mapping is continuous the Picard iterates converge to a fixed point. Details not given here may be found in [3].

To motivate the inwardness aspect of our approach we begin with an example. Let B_r denote a closed ball of radius r centered at the origin in the Banach space X and suppose $t:B_r \to X$ is a contraction mapping with Lipschitz constant $k < 1$. If it is assumed in addition that $T:\partial B_r \to B_r$ (where ∂B_r denotes the boundary of B_r) then Banach's theorem implies existence of a fixed point for T in an indirect way. For, consider the mapping $F = \dfrac{I + T}{2}$ and observe:

(a) F is a contraction mapping.

(b) Any fixed point of F is also fixed under T.

(c) $F:B_r \to B_r$.

We remark that the fact (c) is an easy consequence of the geometry involved: For $x \in B_r$, $x \neq 0$, let $\bar{x} = rx/\|x\|$. Then

$$\|Fx\| = (\tfrac{1}{2})\|x + Tx\| = (\tfrac{1}{2})[\|x - T\bar{x} + Tx\|$$

$$\leq (\tfrac{1}{2})[\|x\| + \|Tx - T\bar{x}\| + \|T\bar{x}\|]$$

$$\leq (\tfrac{1}{2})[\|x\| + k\|x - \bar{x}\| + r] \leq (\tfrac{1}{2})[\|\bar{x}\| + r] = r.$$

By generalizing the above approach Felix Browder obtained the following theorem. (For a proof see [12, page 36]). Notice that in this theorem the fixed points of T_λ and T coincide and if T is a contraction mapping then so is T_λ.

Theorem (Browder). <u>Let</u> K <u>be a bounded convex subset of the Banach space</u> X <u>with</u> int $K \neq \emptyset$ <u>and let</u> T <u>be a Lipschitzian mapping of</u> K <u>into</u> X <u>which carries the boundary of</u> K <u>into</u> K. <u>Then for</u> $\lambda \in (0,1)$ <u>sufficiently small the mapping</u>

$$T_\lambda = (1 - \lambda)I + \lambda T$$

maps K into itself.

This theorem has no metric space counterpart. If (M,d) is a complete metric space which is also metrically convex with B a closed ball in M, and if T:B → M is again a contraction mapping which sends the boundary, ∂B, into B, then it is possible to define the mapping F on B as above by taking Fx always to be a metric midpoint of the pair (x,Tx), x ε B. As before, any fixed point of F will also be a fixed point of T, but neither property (a) nor (c) necessarily holds for F and the preceding approach fails. In this case however, it is possible to define another auxiliary function f whose fixed point will also be a fixed point of T. For x ε B simply define f(x) to be Tx if Tx ε B and let f(x) by any point of ∂B which is metrically between x and Tx if Tx ∉ B. The procedure of Assad-Kirk [1] shows that this mapping f always has a fixed point, and indeed for each x ε B the iterates $\{f^n(x)\}$ always converge to this fixed point.

The assumption that the mapping T maps the boundary of B into B is one of the simplest of the inwardness assumptions, and it turns out to be a much stronger assumption than necessary for existence of fixed points. If T is a contraction mapping defined on a closed subset K of M, then in order to conclude that T has a fixed point it suffices to assume that if x ≠ Tx for x ε K then some point z of M which is metrically between x and Tx with z ≠ x also lies in K. This is the content of Theorem 2 below.

2. Main Result

Our fundamental theorem is a formulation of Caristi's theorem [3] suggested by Felix Browder.

Theorem 1. Let (M,d) be a complete metric space and

$f: M \to M$. If there exists a lower semi-continuous function $\emptyset$ mapping M into the nonnegative real numbers R^+ such that for each $x \in M$

(1) $d(x, f(x)) \leq \emptyset(x) - \emptyset(f(x))$,

then f has a fixed point in M.

Although applications to inwardness situations prompted the discovery of this theorem we call attention to the fact that it is an explicit generalization of Banach's theorem. For if $f: M \to M$ is a contraction mapping, say with Lipschitz constant $k < 1$, then it is easy to see that (1) is satisfied by taking $\emptyset: M \to R^+$ to be:

$$\emptyset(x) = (1 - k)^{-1} d(x, f(x)).$$

Concerning the proof of Theorem 1, condition (1) immediately implies

$$\sum_{i=0}^{n} d(f^i(x), f^{i+1}(x)) \leq \emptyset(x) - \emptyset(f^n(x))$$

and so

$$\sum_{i=0}^{\infty} d(f^i(x), f^{i+1}(x)) < \infty.$$

Thus $\{f^n(x)\}$ is a Cauchy sequence and if f is continuous this sequence must converge to a fixed point of f.

In the general case the procedure is as follows: Let Γ denote the collection of all countable ordinals and select $x_0 \in M$. Using transfinite induction it is possible to define a sequence $\{x_\gamma\}_{\gamma \in \Gamma}$ in such a manner that

(i) $x_\gamma = f(x_{\gamma'})$ if $\gamma = \gamma' + 1$;

(ii) If $\lim_{n \to \infty} \gamma_n = \gamma$ then $\lim_{n \to \infty} x_{\gamma_n} = x_\gamma$.

(For complete details in carrying out the induction consistent with

(ii) see the proof of Theorem 2.1 of Caristi [3].) Next let

$$A = \{x_\gamma : \gamma \in \Gamma\}, \quad m = \inf\{\emptyset(x) : x \in A\}$$

and choose $\{\gamma_i\}$ increasing in Γ so that $\emptyset(x_{\gamma_i}) \to m$ as $i \to \infty$. Since $\lim_{i \to \infty} \gamma_i = \gamma \in \Gamma$, by (ii) $\lim_{i \to \infty} x_{\gamma_i} = x_\gamma$. Lower semicontinuity of $\emptyset$ yields $\emptyset(x_\gamma) \leq m$ and the definition of m with (1) implies $x_\gamma = f(x_\gamma)$.

To see how Theorem 1 implies inwardness results suppose K is a closed subset of the complete metric space M with $T:K \to M$ a contraction mapping. We say that T is <u>metrically inward</u> on K if for each $x \in K$ with $x \neq Tx$ there exists $z \in K$, $z \neq x$, which is metrically between x and Tx (i.e., z satisfies $d(x,z) + d(z,Tx) = d(x,Tx)$). For such mappings it is possible to define $f:K \to K$ so that

$$d(x,f(x)) + d(f(x),Tx) = d(x,Tx), \quad x \in K.$$

(Just take $f(x) = x$ if $x = Tx$; otherwise take $f(x) = z$.) From this,

$$
\begin{aligned}
d(x,f(x)) &= d(x,Tx) - d(f(x),Tx) \\
&\leq d(x,Tx) - [d(f(x),Tf(x)) - d(Tf(x),Tx)] \\
&\leq d(x,Tx) - d(f(x),Tf(x)) + kd(f(x),x)
\end{aligned}
$$

where $k < 1$ is a Lipschitz constant for T. If $\emptyset:K \to R^+$ is defined by

$$\emptyset(x) = (1 - k)^{-1} d(x,Tx), \quad x \in K,$$

then clearly

$$d(x,f(x)) \leq \emptyset(x) - \emptyset(f(x))$$

and Theorem 1 implies:

Theorem 2. Let K be a closed subset of the complete metric space M with $T:K \to M$ a metrically inward contraction mapping. Then T has a fixed point in K.

In certain very restricted circumstances the auxiliary function f associated with $T:K \to M$ of Theorem 2 can be chosen to be continuous and thus have the property that $\{f^n(x)\}$, $x \in K$, always converges to a fixed point.

Theorem 3. Let K be a closed convex subset of a Banach space X with int $k \neq \emptyset$ and suppose $T:K \to X$ is metrically inward and continuous. If the boundary, ∂K, of K contains no segments then the mapping $f:K \to K$ defined by

$$f(x) = \begin{cases} \text{seg}[x,Tx) \cap \partial K, & \text{if } Tx \notin K \\ Tx, & \text{if } Tx \in K \end{cases}$$

is always continuous.

Proof. Suppose $\{x_n\} \subseteq K$ with $x_n \to x$ as $n \to \infty$. By assumption for each n there exists $\alpha_n \in [0,1]$ such that $f(x_n) = \alpha_n x_n + (1 - \alpha_n)Tx_n$, and since $x \in K$ there exists $\alpha \in [0,1]$ such that $f(x) = \alpha x + (1 - \alpha)Tx$. Let $\{\alpha_{n_k}\}_{k=1}^{\infty}$ be any convergent subsequence of $\{\alpha_n\}$, say $\alpha_{n_k} \to \bar{\alpha}$ as $k \to \infty$. Then $f(x_{n_k}) = \alpha_{n_k} x_{n_k} + (1 - \alpha_{n_k})Tx_{n_k} \to \alpha x + (1 - \bar{\alpha})Tx \in K$ as $k \to \infty$ and therefore $\bar{\alpha} \geq \alpha$. If either $\bar{\alpha} = 0$ or $\alpha = 1$ then clearly $\bar{\alpha} = \alpha$. On the other hand if $\bar{\alpha} > 0$ then $Tx_{n_k} \notin K$ for k sufficiently large and thus $\alpha_{n_k} x_{n_k} + (1 - \alpha_{n_k})Tx_{n_k} \in \partial K$ and either $\bar{\alpha} = \alpha$ or $\bar{\alpha} = 1$. But by assumption ∂K contains no segments so if $\alpha < 1$ then there exists $\lambda \in (\alpha, 1)$ such that $\lambda x + (1 - \lambda)Tx \in$ int K; thus for k sufficiently large $\lambda x_{n_k} + (1 - \lambda)Tx_{n_k} \in$ int K and $\alpha_{n_k} \leq \lambda$ implying $\bar{\alpha} \leq \lambda < 1$. Therefore in any case $\bar{\alpha} = \alpha$ and it follows that $\{\alpha_n\}$ converges to α. Hence $f(x_n) = \alpha_n x_n + (1 - \alpha_n)Tx_n \to \alpha x + (1 - \alpha)Tx = f(x)$ as $n \to \infty$, completing the proof.

3. Inwardness Results in Banach Spaces

Now assume X is a Banach space, $K \subset X$, and define the inward set $I_K(x)$ of $x \in X$ with respect to K as:

$$I_K(x) = \{x + c(u - x) : u \in K, \ c \geq 0\}.$$

A mapping $T:K \to X$ is called inward if $Tx \in I_K(x)$ for all $x \in K$. If $Tx \in \overline{I_K(x)}$ for $x \in K$ then T is said to be weakly inward. These definitions are due to Halpern [4].

Using notation $d(z,K) = \inf\{\|z - x\| : x \in K\}$ we have:

Theorem 4. Suppose $K \subset X$ is convex and let $f:K \to X$. Then the mapping $T = I - f$ is weakly inward on K if and only if for each $x \in K$,

$$(2) \quad \lim_{h \to 0^+} \frac{d(x - hf(x), K)}{h} = 0.$$

H. Brezis proved in 1970 [2] that (2) is equivalent to flow invariance for the differential equation $x' = f(x)$ in R^n (where $x:[0,\infty) \to R^n$ and $f:K \subset R^n \to R^n$ is locally Lipschitzian).

Combining the characterization of Theorem 4 with Theorem 1 we obtain the following two theorems. These theorems are also contained, respectively, in papers of Martin [6] and Vidossich [11] where condition (2) is used rather than explicit reference to inwardness.

Theorem 5. [3,6] Let X be a Banach space, K a closed convex subset of X, and $T:K \to X$ a weakly inward contraction mapping. Then T has a fixed point in K.

Theorem 6. [3,11] Let X be a Banach space, and suppose K is a closed convex subset of X which possesses the fixed point property with respect to nonexpansive self-mappings. Suppose $T:K \to X$ is a nonexpansive (or more generally, a Lipschitzian pseudo contractive) mapping which is weakly inward. Then T has a fixed point in K.

Theorem 5 for the more general 'condensing' mappings but with the assumption that T is inward rather than weakly inward and also Theorem 6 with the stronger inwardness assumption appear respectively in two papers of Reich [9,8].

For an indication of the proof of Theorem 5, suppose $T:K \to X$ is a weakly inward contraction mapping. Let $k < 1$ be a Lipschitz constant for T. Choose $\epsilon > 0$ so that $k < (1 - \epsilon)/(1 + \epsilon)$. If $x \in K$ with $x \neq Tx$ then by Theorem 4 there exists $h \in (0,1)$ such that, with $T = I - f$,

$$d((1 - h(x + hTx, K) < h \ \epsilon \ \|x - Tx\|.$$

Next select $y \in K$ so that

(3) $\|\bar{x} - y\| < h \ \epsilon \ \|x - Tx\|$

where $\bar{x} = (1 - h)x + hTx$. For this y it is possible to show that

$$\|y - Ty\| < \|x - Tx\| + (k - \frac{1 - \epsilon}{1 + \epsilon}) \ \|x - y\|.$$

Define $f:K \to K$ by $y = f(x)$. Then for $x \in K$

$$\|x - f(x)\| \leq (\frac{1 - \epsilon}{1 + \epsilon} - k)^{-1} \ [\|x - Tx\| - \|f(x) - Tf(x)\|]$$

where $x = f(x)$ if and only if $x = Tx$. By Theorem 1 (with $\emptyset(x) = (\frac{1 - \epsilon}{1 + \epsilon} - k)^{-1} \|x - Tx\|$) f has a fixed point.

REFERENCES

[1] N.A. Assad and W.A. Kirk, Fixed point theorems for set valued
     mappings of contractive type, Pacific J. Math. 43 (1972),
     553-562.

[2] H. Brezis, On a characterization of flow invariant sets, Comm.
     Pure Appl. Math. 23 (1970), 261-263.

[3] J. Caristi, Fixed point theorems for mappings satisfying in-
     wardness conditions.

[4] B. Halpern, Fixed point theorems for outward maps, Doctoral
     Thesis, U.C.L.A. (1965).

[5] B. Halpern and G. Bergman, A fixed point theorem for inward
     and outward maps, Trans. Amer. Math. Soc. 130 (1968), 353-358.

[6] R.H. Martin, Differential equations on closed subsets of a
     Banach space, Trans. Amer. Math. Soc. 179 (1973), 399-414.

[7] W.V. Petryshyn and P.M. Fitzpatrick, Fixed point theorems for
     multivalued noncompact inward maps.

[8] S. Reich, Remarks on fixed points, Atti Accad. Naz. Lincei
     Rend. Cl. Sci. Fis. Mat. Natur. 52 (1972), 690-697.

[9] S. Reich, Fixed points of condensing functions, J. Math. Anal.
     Appl. 41 (1973), 460-467.

[10] S. Reich, Fixed points of nonexpansive functions, J. London
     Math. Soc. 7 (1973), 5-10.

[11] G. Vidossich, Nonexistence of periodic solutions of differen-
     tial equations and applications to zeros of nonlinear opera-
     tors.

[12] G. Vidossich, Applications of topology to analysis: On the
     topological properties of the set of fixed points of nonlinear
     operators, Confer. Sem. Mat. Univ. Bari 126 (1971), 1-62.

ON SOME ASPECTS OF FIXED POINT THEORY IN BANACH SPACES

Michael Edelstein
Dalhousie University

ABSTRACT

Let $T:X \to X$ be an affine mapping of the Banach space X into itself satisfying the condition

$$\|T^n x - T^n y\| \leq c\|x - y\| \quad (x,y \in X; \quad n = 1,2,\ldots)$$

for some $c > 0$. If an $x \in X$ exists such that the sequence $\{x_n\}$ where $x_n = \frac{1}{n} \sum_1^n T^k x$, contains a subsequence $\{x_{n_i}\}$ which converges weakly to an $\bar{x} \in X$ then $T\bar{x} = \bar{x}$ and, moreover, $\{x_n\}$ converges strongly to $\bar{x}$.

If $f:X \to X$ is a nonexpansive mapping of a strictly convex Banach space into itself such that, for some $y \in X$, x is a cluster point of the sequence $\{f^n(y)\}$ of iterates then the restriction of f to $\overline{co}\{f^n(x)\}$ is known to be an affine isometry. Using this and the fact that such an isometry can be extended to an affine isometry of the affine space of $\overline{co}\{f^n(x)\}$ into itself one derives the following result:

With f and X as above if $\{\frac{1}{n} \sum_1^n f^k(x)\}$ contains a subsequence which converges weakly to some ξ then $f(\xi) = \xi$ and the above sequence converges strongly to ξ.

ON SOME ASPECTS OF FIXED POINT THEORY IN BANACH SPACES

Michael Edelstein
Dalhousie University

In recent works on fixed point theory much attention is being given to special classes of mappings, e.g. the nonexpansive mappings, the isometries and the affine mappings. We intend to dwell here on interconnections between the above mentioned types of mappings. To begin with we recall three well known results in the area.

I. (Markov-Kakutani [8],[10]): Each family of commuting affine continuous mappings of a compact convex set, in a locally convex Hausdorff space, into itself, has a common fixed point.

II. (Brodski-Milman [1]): Every weakly compact convex subset K of a Banach space such that K has normal structure contains a point c which is fixed under all isometries of K onto itself.

III. (Kirk [9]): With K as above, if $f:K \to K$ is nonexpansive then f has a fixed point (in K).

These results have motivated a great deal of research in fixed point theory. For example, the first led to considerable efforts which have been directed towards the replacement of the commuting property with some structural property of the family. The second is more geometrical in nature in that the property of normal structure is used to construct a "center" c with an a priori fixed point property for the class of mappings in question. (Recall that a convex set K is said to have normal structure if every nonempty bounded convex subset C, which is not a singleton contains a point x such that $\sup\{\|x - y\|:y \in C\}$ is smaller than the diameter of C.) Kirk's use of normal structure in his result has led to further work of this and related geometric properties of Banach spaces. The interest in nonexpansive mappings was further

increased by the simultaneous appearance of results of Browder [2]
and Göhde [6] which are essentially special cases of Kirk's theorem
(in that the Banach space is assumed uniformly convex, so that each
convex subset has normal structure).

Nonexpansive mappings of a Banach space into itself have, in
general, little in common with isometries or, for that matter, with
affine mappings. However under the additional assumptions given
below the connection becomes quite strong.

(a) In any metric space (X,d), if $f:X \to X$ is nonexpansive
and x is a cluster point of a sequence of iterates $\{f^n(y)\}$ then
$\{f^n(x)\}$ has the property that the restriction of f to it is an
isometry (cf [4]).

(b) If X is a strictly convex Banach space and f and x
are as before then the restriction of f to the closed convex hull
of $\{f^n(x)\}$ is an affine isometry (cf [5]).

An observation which makes (b) noteworthy is [5] that any
affine isometry of a convex set C in a normed linear space can be
extended to such an isometry of the closed affine hull of that set,
mapping it into itself. Thus if $L(C)$ is the affine hull of C
where $C = \overline{co}\{f^n(x)\}$ there is an affine isometry $h:L(C) \to L(C)$
such that $h|C = f|C$.

It follows that if h has a fixed point $z \in L(C)$ then the
(unique) nearest point w in C must be fixed under f. Here
then the problem of finding fixed points for f is reduced to one
of finding such points for affine isometries. It so happens that
no greater effort is required for dealing with the more general
class of affine mappings which satisfy the condition

(1) $\|T^n x - T^n y\| \leq c\|x - y\|$ $(x,y \in X;\quad n = 1, 2, \dots)$

for some $c > 0$.

Writing $Tx = Ax + a$ with $a = T0$, this is equivalent to
requiring that the linear operator A be power bounded, i.e.

$\|A^n\| \leq c$ $(n = 1,2,\ldots)$. (Indeed $T^n x = A^n x + A^{n-1} z + \ldots + Aa + a$ so that $T^n x - T^n y = A^n (x - y) \Rightarrow \|T^n x - T^n y\| = \|A^n (x - y)\|$.) For such mappings we have the following.

Theorem A. Let T be an affine mapping of a Banach space X into itself satisfying condition (1) and such that for some $x \in X$ the sequence $\{x_n\}$, where $x_n = \frac{1}{n} \sum_1^n T^k x$, contains a subsequence $\{x_{n_i}\}$ which converges weakly to an $\bar{x} \in X$. Then $T\bar{x} = \bar{x}$ and $\{x_n\}$ converges strongly to $\bar{x}$.

Proof. Set $Tx = Ax + a$ where $a = T0$ and A is linear. Then A is power bounded with $\|A^n\| \leq c$, and therefore, $\frac{1}{n}\|A^{n+1} x - Ax\| \to 0$ as $n \to \infty$. Now, by direct computation one obtains that $(A - I)x_n = \frac{1}{n}(A^{n+1} x - Ax) + \frac{1}{n} \sum_1^n A^k a - a$ where I stands for the identity mapping on X. Since $A - I$ is weakly continuous it follows that $(A - I)x_{n_i} \longrightarrow (A - I)\bar{x}$ and, therefore, $\frac{1}{n} \sum_I^{n_i} A^k a$ converges weakly to some element of X, say $\bar{a}$. By the Kakutani-Yosida mean ergodic theorem [7] we have $A\bar{a} = \bar{a}$ and, moreover, $\frac{1}{n} \sum_1^n A^k a$ converges strongly to $\bar{a}$. To prove that $T\bar{x} = \bar{x}$ it clearly suffices to show that $\bar{a} = 0$. Now, from $(A - I)\bar{x} = \bar{a} - a$ it readily follows that

$$T^k \bar{x} = \bar{x} + k\bar{a} \quad (k = 1,2,\ldots)$$

so that

$$\frac{1}{n} \sum_1^n T^k x = \frac{1}{n} \sum_1^n (\bar{x} + k\bar{a}) + \frac{1}{n} \sum_1^n (T^k x - (\bar{x} + k\bar{a}))$$

$$= \bar{x} + \frac{n + 1}{2} \bar{a} + \frac{1}{n} \sum_1^n (T^k x - T^k \bar{x})$$

and

$$\left\|\frac{1}{n} \sum_1^n T^k x\right\| \geq \left\|\bar{x} + \frac{n + 1}{2} \bar{a}\right\| - \frac{1}{n}\left\|\sum_1^n (T^k x - T^k \bar{x})\right\|$$

$$\geq \left\|\bar{x} + \frac{n + 1}{2} \bar{a}\right\| - c\|x - \bar{x}\|.$$

If $\bar{a} \neq 0$ then both sides of the last inequality must tend to infinity as $n \to \infty$, which is incompatible with the assumption that $x_{n_i} \longrightarrow \bar{x}$. Thus $\bar{a} = 0$ and $T\bar{x} = x$ to complete the proof consider the identities

$$\frac{1}{n} \sum_1^n A^k x = \frac{1}{n} \sum_1^n T^k x - \frac{1}{n} \sum_1^n T^k 0$$

and

$$\frac{1}{n} \sum_1^n A^k \bar{x} = \frac{1}{n} \sum_1^n T^k \bar{x} - \frac{1}{n} \sum_1^n T^k 0 = \bar{x} - \frac{1}{n} \sum_1^n T^k 0.$$

By subtraction we obtain

$$\frac{1}{n} \sum_1^n A^k (x - \bar{x}) = \frac{1}{n} \sum_1^n T^k x - \bar{x}$$

Since $x_{n_i} - \bar{x} \longrightarrow 0$ it follows that $\frac{1}{n} \sum_1^{n_i} A^k (x - \bar{x}) \longrightarrow 0$ so that the Kakutani-Yosida Theorem applies again to the effect that $\frac{1}{n} \sum_1^n A^k (x - \bar{x})$ converges strongly to 0. It follows that the entire sequence $\{x_n\}$ converges strongly to $\bar{x}$ as claimed.

Remark. A similar result was stated in [5]. However, the proof given there seems to be inadequate.

From Theorem A we can easily derive the following.

Theorem B. Let X be a strictly convex Banach space and $f:X \to X$ a nonexpansive mapping. If an $x \in X$ exists such that x is a cluster point of a sequence of iterates and $\{\frac{1}{n} \sum_1^n f^k(x)\}$ contains a subsequence which converges weakly to $\xi \in X$ then $f(\xi) = \xi$ and $\{\frac{1}{n} \sum_1^n f^k(x)\}$ converges strongly to ξ.

In the case of a reflexive Banach space X, if T is affine and satisfies (1) the boundedness of any sequence of iterates $\{T^n x\}$ clearly implies the boundedness of the sequence $\{x_n\}$ and, therefore, the existence of a weakly convergent subsequence. Thus in the preceding theorems we may replace the assumption of arithmetic means with a boundedness assumption on a sequence of iterates.

It is of some interest to note that for X finite dimensional (and strictly convex) the assumption on the existence of a cluster point for some sequence of iterates suffices to guarantee the existence of a fixed point so that the other assumption becomes redundant. For infinite dimensional, even separable Hilbert spaces one can show by suitable examples (cf [5]) that this is no longer the case. The question whether strict convexity is necessary or not in the finite dimensional case is still open.

In closing we would like to mention the following application of Theorem A to linear functional equations in Banach spaces.

Theorem C. Let X be a Banach space, A a power bounded linear operator. For the equation

$$x = Ax + a$$

to have a solution it suffices that the sequence $\{y_n\}$, where $y_n = \frac{1}{n} \sum_1^n T^k x$ $Tx = Ax + a$, have a weakly convergent subsequence. Moreover, when this is the case $\{y_n\}$ converges strongly to a solution.

The last theorem which is an immediate consequence of Theorem A improves a corresponding result ([3], p.22, Theorem 1.8) of de Figueiredo in that it removes as unnecessary the hypothesis that $\{\|\sum_1^n A^k a\| : n = 1, 2, \ldots\}$ be bounded.

REFERENCES

[1] M.S. Brodski, D.P. Milman, On the center of a convex set (Russian), Dokl. Akad. Nauk SSSR 59 (1948), 837-840.

[2] F.E. Browder, Nonexpansive nonlinear operators in a Banach space, Proc. Nat. Acad. Sci. 54 (1965), 1041-1044.

[3] D.G. de Figueiredo, Topics in nonlinear functional analysis, University of Maryland, Lecture Series, No. 48, pp.1-53.

[4] M. Edelstein, On nonexpansive mappings, Proc. Am. Math. Soc. 15 (1964), 689-695.

[5] M. Edelstein, On nonexpansive mappings of Banach spaces, Proc. Cambr. Phil. Soc. 60 (1964), 439-447.

[6] D. Göhde, Zum Prinzip der kontraktiven Abbildung, Math. Nach. 30 (1965), 251-258.

[7] S. Kakutani, K. Yosida, Operator theoretical treatment of Markov's process and mean ergodic theorem, Ann. of Math. 42 (1941), 188-228.

[8] S. Kakutani, Two fixed point theorems concerning bicompact convex sets, Proc. Imp. Acad. Tokyo 14 (1938), 242-245.

[9] W.A. Kirk, A fixed point theorem for mappings which do not increase distance, Am. Math. Monthly 72 (1965), 1004-1006.

[10] A. Markov, Quelques théorèmes sur les ensembles abeliens, Dokl. Akad. Nauk SSSR 10 (1936), 311-314.

MIMICRY IN NORMED SPACES

Mahlon M. Day
University of Illinois

§0. Introduction

A partial order relation N mimics C will be defined in the family of normed spaces and a superreflexive space will be defined to be a space which mimics no non-reflexive space. The purpose of this talk is to describe the work of R.C. James and of Enflo which characterizes superreflexive spaces in terms of geometric properties of the unit ball. The properties I shall discuss are uniform rotundity, uniform non-squareness, and a property best described by saying that all ϵ-trees grow as the number of branches is increased. For general reference to notation see my book Day 1973.

§1. Mimicry and Superreflexivity

Definition. Say that N **mimics** C [symbol: $N > C$] [James 1972 says "C is finitely representable in N"] whenever each finite-dimensional subspace E of C is arbitrarily nearly isometric with a subspace of N; that is, there is a linear, one-to-one function T from E into N such that $\|T\| \cdot \|T^{-1}\|$ is as near 1 as is desired.

(1a) **Example.** For each normed space N the space c_0 mimics N. The proof depends on the presence of cubes of large finite dimension as the unit balls of subspaces of c_0. Illustrating with $E = \ell^2_2$, it is clear that a polygon arbitrarily near the circle can be made with enough lines $f_i^{-1}(\pm 1)$, with $\|f_i\| = 1$ and $i = 1,\ldots,n$, so that the map $TX = (f_1(x),\ldots,f_n(X),0,0,\ldots)$ can be made as near isometric as desired by taking n large. This shows that <u>a non-reflexive space can mimic a reflexive space</u>. <u>The reverse is also true</u>.

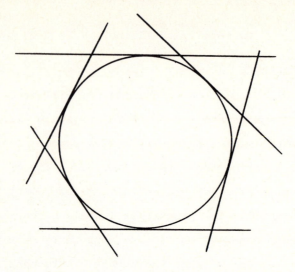

(1b) <u>Example</u>. Take $1 < p < \infty$ and let P be the substi-
tution space $\ell^p m_n$, where m_n is the n-dimensional space with
cube for unit ball and P is the space of sequences $\tilde{x} = (x_n)$
where $x_n \in m_n$ and $\|\tilde{x}\|_p = (\sum_{n \in \omega} (\|x_n\|_m)^p)^{1/p} < \infty$. Then P <u>is</u>
<u>reflexive and if</u> N <u>is a normed space</u>, P <u>mimics</u> N.

Definitions. N is ρ means that the completion of N is
reflexive. N is <u>superreflexive</u> [N is $\sum_\rho$] means that if N
mimics C then C is ρ.

(1c) All finite-dimensional normed spaces are superreflexive.
Indeed, in FD, the class of finite-dimensional spaces, $N > C$ if
and only if C is linearly isometric with a subspace of N. Obviously, if C is isometric with a subspace of N, then $N > C$
no matter what the dimensions of either.

(1d) $>$ is reflexive and transitive but not symmetric nor
antisymmetric.

(1e) Each superreflexive space is reflexive.

(1f)   For each normed space   N   it follows that   N** > N.
The converse was proved by Lindenstrauss and Rosenthal 1969.

(1g)   If   N   is a normed space, then   N > N**.

Among the infinite-dimensional normed spaces there is a minimal one.   This result is due to Dvoretsky 1961; V.D. Mil'man 1971,
Figiel 1972, and Szankowski 1974 have presented other versions of
the proof but no simple proof is known.

(1h)   Every infinite-dimensional normed space   N   mimics ℓ^2.

Notes.   (1i) ℓ^1 does not mimic m_3.   [James, quoted in Day
1973, page 173.]   (1j)   There is a non-reflexive space which does
not mimic ℓ^1_3.   [James, to appear.]   (1k) ℓ^2 mimics all inner-
product spaces and no other spaces.   [Use the Jordan – von Neumann
characterization of inner-product spaces, for example from Day
1973, page 151.]

Here is a blob diagram for the partially ordered system of all
normed spaces with the relation   >.

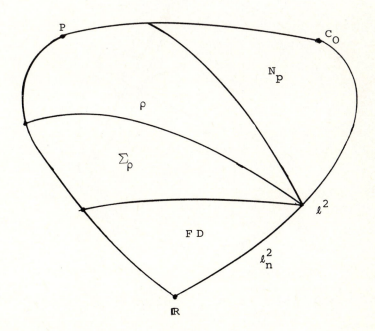

§2. Some Geometric Properties of Normed Spaces

Here $U = U(N)$ is the unit ball of N and $\Sigma = \Sigma(N)$ is the surface of U.

Definition. N is uniformly rotund, UR, [Clarkson 1936 calls it uniformly convex] means that for each $\epsilon > 0$ there is $\delta(\epsilon) > 0$ such that whenever x and y are elements of $U(N)$ and $\|x - y\| \geq \epsilon$ then $(x + y)/2$, the midpoint of the segment from x to y, is at depth at least $\delta(\epsilon)$ below Σ.

Examples of Clarkson are the spaces ℓ^p and L^p with $1 < p < \infty$.

Definitions. N is quadrate, Q, [Beck 1962 calls it not uniformly non-square] means that for each $\delta > 0$ there are x and y in $U(N)$ such that $\|x + y\| > 2(1 - \delta)$ and $\|x - y\| > 2(1 - \delta)$. N is inquadrate, NQ, if N is not quadrate.

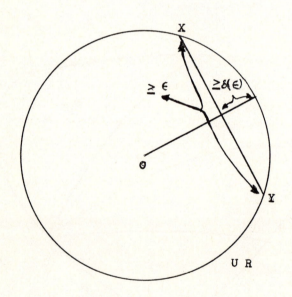

(2b) [Beck 1962] If N is uniformly rotund, then N is inquadrate.

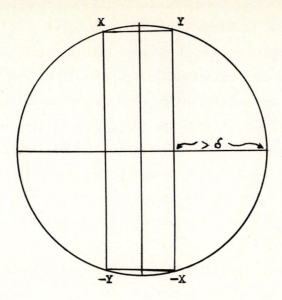

<u>Trees: Pictures</u>

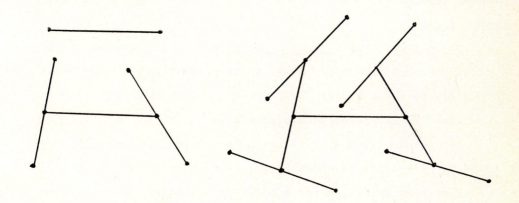

<u>Trees: Formal description</u>.

Let Φ be the set of all ultimately zero real sequences, let
b_i be the usual basis vector with   1   in place   i   and   0   elsewhere, and let ϵ_i be   <u>+</u>1.

The <u>basic</u> 1-<u>tree</u> is the pair of 1-<u>twigs</u> $\epsilon_1 b_1$;   that is, b_1
and $-b_1$.

The <u>basic</u> 2-<u>tree</u> is the basic 1-tree supplemented by the two pairs of 2-<u>twigs</u> $b_1 + \epsilon_2 b_2$ and $-b_1 + \epsilon_3 b_3$.

When the basic n-tree is defined with its 2^{n-1} pairs of n-twigs built out of combinations of the first $2^n - 1$ basis vectors arranged, say, by lexicographic order of the sequences of coefficients as $x_1, \ldots, x_{2^n}$, then the <u>basic</u> (n + 1)-<u>tree</u> is the basic n-tree supplemented by the 2^n pairs of (n + 1)-twigs $x_i + \epsilon_{k_i} b_{k_i}$ where $k_i = i - 1 + 2^n$. For example, the four pairs of 3-twigs of the basic 3-tree are $b_1 + b_2 + \epsilon_4 b_4$, $b_1 - b_2 + \epsilon_5 b_5$, $-b_1 + b_2 + \epsilon_6 b_6$, and $-b_1 - b_2 + \epsilon_7 b_7$. The <u>basic</u> ω-<u>tree</u> is the union of all the basic n-trees.

An $n - \epsilon$-tree in a normed space N is a translate of a linear image of the basic n-tree such that the tips of each matched pair of i-twigs, $1 \leq i \leq n$, are at least ϵ apart. [Or, if the linear mapping from Φ into N is T, then $\|Tb_i\| \geq \epsilon/2$ for each $i < 2^n$.

<u>Definition</u>. <u>All ϵ-trees grow in</u> N, N is Y, means that for each $\epsilon > 0$ there is $n(\epsilon)$ so large that no $n - \epsilon$-tree with $n \geq n(\epsilon)$ is contained in $U(N)$.

(2c) If N is uniformly rotund, then N is Y.

[In passing from (i + 1)-twigs to i-twigs, the norm is multiplied by $1 - \delta(\epsilon)$; eventually the i-twigs could not be ϵ long.]

(2d) Let $r(n, \epsilon) = \inf r > 0 | rU(N)$ contain an $n - \epsilon$-tree. Then $r(n, \epsilon) \leq n\epsilon/2$ and if N is Y, then $\lim_n r(n, \epsilon) = \infty$.

[For if no $n - \epsilon/r$-tree fits in U, then no $n - \epsilon$-tree fits in rU.]

This accounts for my name for this property Y and also for the next result.

(2e) Y (like ρ) is isomorphism invariant.

Note that UR and Q are not isomorphism invariant; example: ℓ^2_2 is isomorphic to ℓ^1_2.

Notation. If π is some property of normed spaces we shall write N is $\langle\pi\rangle$ to mean that N is isomorphic to some space M for which M is π. We shall also say that N is Nπ when N fails to be π. (2e) can now be rephrased as $\langle\gamma\rangle = \gamma$ and $\langle\rho\rangle = \rho$.

Our Goal For Today

To discuss the proof of equivalence of

$$\Sigma\rho, \langle\Sigma\rho\rangle, \gamma, \langle UR\rangle, \langle NQ\rangle,$$

and one or two other conditions if time should permit.

(2f) [Day 1941] The space P of (1b) is reflexive but not $\langle UR\rangle$.

We have a start on our goal already.

$$\gamma \overset{2a}{\Longleftarrow} \langle UR\rangle \overset{2b}{\Longrightarrow} \langle NQ\rangle$$

$$2a \;\Big\Downarrow\;\!\!\nwarrow\; 2f$$

$$\rho$$

§3. The First Hard Steps

James is responsible for the work in sections 3, 4 and 5. Beck asked if NQ implies ρ. James 1964 shifted to the contra-positive and proved it.

(3a) N_ρ implies γ.

James' extremely ingenious proof turned out to generalize to other circumstances. Schäffer and Sundaresan 1968 use the method to study the girth of spheres and its relation to reflexivity, and James 1972 adapted the method to show the next necessary fact.

(3b) In each normed space with non-reflexive completion there is an $\epsilon > 0$ for which ϵ-trees need not grow; that is N_ρ implies N_γ.

The diagram is now much improved; the inner heavy arrows are merely the contrapositives of the outer ones.

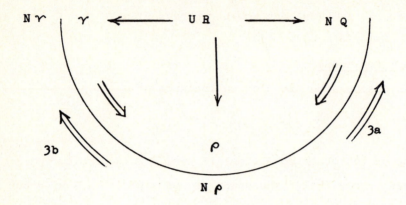

§4. <u>Stability of Properties Under Mimicry</u>

Next consider how properties of N itself relate to properties of <u>at least one</u> C such that N mimics C.

(4a) If N mimics C and if N is UR or NQ or γ, so
is C.

(4b) Every uniformly rotund space is superreflexive; that is,
UR implies $\Sigma\rho$.

[(4a) and N mimics C give that C is UR; then (2a) implies that C is ρ. Hence N is $\Sigma\rho$.]

(4c) If N mimics C and N is superreflexive, so is C.

[By transitivity of mimicry. Recall that no implication holds
between ρ for N and ρ for C.]

In the next diagram properties on the inner ring are properties of N and those on the outer ring are properties of <u>at least
one</u> C such that N mimics C.

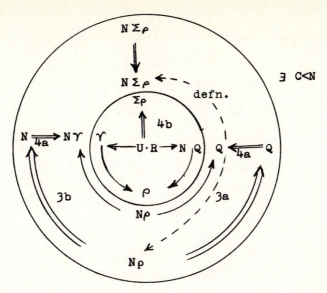

Now trace arrows starting from $N\Sigma\rho$ for the space N to N_ρ for C.

(4d) If N is not superreflexive, then N is both Q and N. See the diagram below.

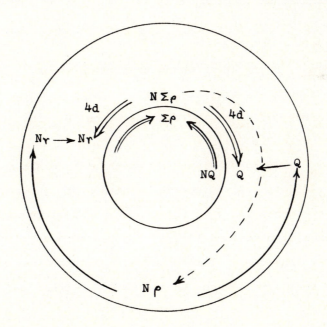

The usual shift to contrapositives again increases the information in the inner diagram of isomorphism classes for N.

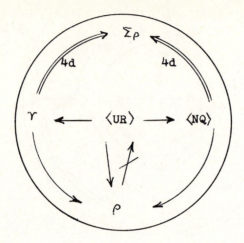

To close up the cycle we still need two results:

From section 5: <u>Trees grow in superreflexive spaces</u>; that is, N_γ implies $N\Sigma\rho$. [James 1972]

From section 6: <u>If trees grow in N, then N can be renormed to be uniformly rotund</u>; that is, γ implies $\langle UR\rangle$. [Enflo 1972]

§5. <u>Tree Growth</u>

<u>Definition</u>. N is N_{γ_ω} means that there is an $\epsilon > 0$ for which there is an $\omega - \epsilon$-tree in $U(N)$.

(5a) [James 1972] If N is N_{γ}, then N mimics at least one space C which is N_{γ_ω}.

[Take T_n mapping Φ into N so that the basic n-tree goes into an $n - \epsilon$-tree in U. Since the elements of Φ with rational coordinates are dense in c_o, the Cantor diagonal process applied to the sequence of functions $(\|T_n x\|)$ gives a convergent subsequence (T_{n_i}). The limit $\lim_i \|T_{n_i} z\| = /x/$ determines a norm in Φ such that the basic ω-tree becomes an $\omega - \epsilon$-tree and N mimics $(\Phi, / /)$.]

(5b) If C is N_{γ_ω}, then C is N_ρ and hence C is also Q. [James 1972]

(5c) N_γ implies $N\Sigma\rho$ and Q. [James 1972]

[By (5a) N mimics some C with N_{γ_ω}; by (5b) C is Q and N_ρ; hence N is $N\Sigma\rho$ and (by (3a)) is Q.]

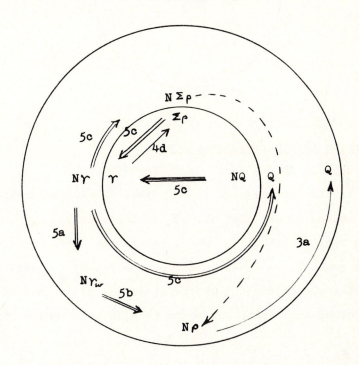

Hence our inner isomorphism-class diagram now shows that γ and $\Sigma\rho$ are equal [so, by (2e), $\Sigma\rho$ is isomorphism invariant, although the definition did not show it clearly] and both γ and $\Sigma\rho$ are implied by <UR> and <NQ>.

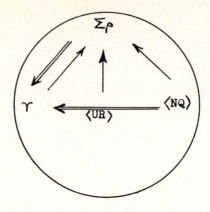

Here endeth the part done by James.

§6. Renorming the Uniform Rotundity

This depends on the paper of Enflo 1972.

(6) If all ϵ-trees grow in N, then N is isomorphic to a uniformly rotund space; that is, Υ implies $\langle UR \rangle$.

Enflo begins with 2^{-n}, where n is the $n(2, \epsilon)$ from (2d) as a first guess for a $\delta(\epsilon)$ and, for one fixed ϵ, gets a function that has midpoints of ϵ-chords at depth $\delta(\epsilon)$. Working with combinations of such functions he is able to construct a UR norm in N.

[A manuscript of Pisier (to appear) observes that trees are interpretable as dyadic martingales and that Enflo's construction can be replaced by one using martingales more general than those used in Enflo's proof. Pisier concludes that in a space N with Υ there is a p with $2 \leq p < \infty$ such that N can be given a new norm in which the modulus of rotundity of the new norm is of pth power type, that is, there is a positive number η such that if $\|x\| \leq 1$ and $\|y\| \leq 1$, then $\|(x + y)/2\| \leq 1 - \eta \|x - y\|^p$.]

This is enough to complete the equivalence proof.

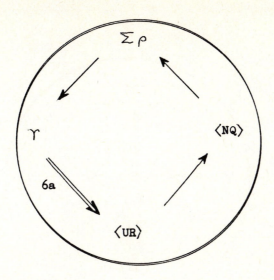

§7. <u>Duality and Smoothness</u>

Many years ago I observed that the familiar examples of uniformly smooth spaces, that is, the ℓ^p and L^p spaces with $1 < p < \infty$, were uniformly rotund as well as smooth, and I asked whether $\langle US \rangle = \langle UR \rangle$? Enflo showed this to be true.

(7a) N is uniformly rotund if and only if N* is uniformly smooth [called uniformly flattened in Day 1944] and if and only if N* has norm which is uniformly Fréchet differentiable on the unit sphere [Šmul'yan 1938].

(7b) If N is Q, so is N*. [James 1964]

(7c) A normed space N is $\langle NQ \rangle$ if and only if N* is $\langle NQ \rangle$.

[For if N or N* is NQ, N is ρ; use (7b). [James 1964]]

(7d) $\Sigma\rho$, γ, $\langle UR \rangle$, and $\langle NQ \rangle$ hold in N if and only if they hold in N*. [By (7a) and §6.]

(7e) N is $\langle UR \rangle$ if and only if N is $\langle US \rangle$ if and only if N* is $\langle UR \rangle$. [By (7a) and (7d).]

Adding these to the diagram at the end of §6 gives the final result.   [Enflo 1972]

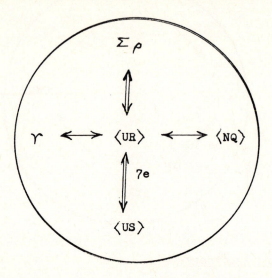

Of course the work of Asplund 1967 on averaged norms now shows that

(7f)   N   can be given a single norm under which   N   and   N*
are both   UR   and   US.

BIBLIOGRAPHY

General reference: Day, M.M. (1973) Normed linear spaces, 3rd edition. Ergeb. der Math. vol. 21, Springer-Verlag [especially Chapter VII, section 4B.]

Asplund, E. (1967) Averaged norms, Israel Jour. of Math. 5, 227-233.

Beck, A. (1962) A convexity condition in Banach spaces and the strong law of large numbers, Proc. Amer. Math. Soc. 13, 329-344.

Clarkson, J.A. (1936) Uniformly convex spaces, Trans. Amer. Math. Soc. 40, 396-414.

Day, M.M. (1941) Reflexive Banach spaces not isomorphic to uniformly convex spaces, Bull. Amer. Math. Soc. 47, 313-317.

_____ (1944) Uniform convexity in factor and conjugate spaces, Ann. of Math. (2) 45, 375-385.

Dean, D.W. (1973) The equation $L(E,X^{**}) = L(E,X)^{**}$ and the principle of local reflexivity, Proc. Amer. Math. Soc. 40, 146-148.

Dvoretsky, A. (1961) Some results on convex bodies and Banach spaces, Proc. Internat. Sympos. Linear spaces, July 1960, Jerusalem, Hebrew University of Jerusalem 1961.

Enflo, P. (1972) Banach spaces which can be given an equivalent uniformly convex norm, Israel Jour. of Math. 13, 281-288.

Figiel, T. (1972) Some remarks on Dvoretsky's theorem on almost spherical sections of convex bodies, Colloq. Math. 24, 241-252.

James, R.C. (1964) Uniformly non-square Banach spaces, Ann. of Math. 80, 542-550.

_____ (1972) Some self-dual properties of normed linear spaces Symposium on Infinite-dimensional Topology, Ann. of Math. Studies 69, 159-175.

_____ (to appear) A non-reflexive Banach space that is uniformly non-octahedral. (Preprint dated 3/1/74.)

Lindenstrauss, J. and Rosenthal, H.P. (1969) The L_p spaces, Israel Jour. of Math. 7, 325-349.

Mil'man, D.P. (1938) On some criteria for the regularity of spaces of the type (B). C.R. (Doklady) Acad. Sci. USSR 20, 243-246.

Mil'man, V.D. (1971) A new proof of the theorem of A. Dvoretsky on sections of convex bodies. Funkcional. Anal. i Prilozen 5, 28-37. (Also in the complete English translation of this journal.)

Pettis, B.J. (1939) A proof that every uniformly convex space is reflexive, Duke Math. Jour. 5, 249-253.

Schäffer, J.J. and Sundaresan, K. (1968) Reflexivity and the girth of spheres, Math. Ann. 184, 169-171.

Smul'yan, V. (1940) Sur la derivabilité de la norm dans l'espace de Banach, C.R. (Doklady) Acad. Sci. USSR 27, 643-648.

Szankowski, A. (1974) On Dvoretsky's theorem on almost spherical sections of convex bodies, Israel Jour. of Math. 17, 325-338.

SMOOTHNESS CLASSIFICATION OF REFLEXIVE SPACES I

Kondagunta Sundaresan*
University of Wyoming

In this paper a method of classifying reflexive real Banach space upto isometry based on higher order differentials of the norm is provided. Since the boundary of the unit cell in a Banach space is a level surface of the norm functional of the space the differentials of the norm are useful in the study of the geometry of the cell. Thus the classification of reflexive spaces provided here is in a sense geometric. The classification is further extended so that if two Banach spaces are isomorphic they belong to the same class.

The paper is divided into three sections. In Section 1 basic definitions and notations are stated for convenience of reference. The classification of reflexive spaces, and illustrative examples are provided in Section 2. In Section 3 curvature of the boundary of the unit cell is studied, thus pointing out the importance of the classification in the geometry of the surface of the unit ball.

Section 1. Basic Definitions and Notations

It is known that the classical differential calculus which is useful in the smoothness study of curves has its natural counter-part in Banach spaces. For a vivid account of differential calculus in Banach spaces, the reader is referred to Dieudonne [7] and Cartan [3].

If E is a Banach space U_E, S_E denote respectively the unit cell and the boundary of the unit cell respectively. If E is understood these are simply denoted by U and S respectively. A closed convex centrally symmetric set in E, with non-empty interior is called a convex body here. E^* denotes the dual of the

---

*The research was in part supported by NSF Grant GP-43213.

space   E,   and U^*, S^* are respectively the unit cell and the unit
sphere of E^* if   E   is understood.

If   E,F   are two Banach spaces the Banach space of continuous
linear operators on   E → F   is denoted by   L(E,F).   The Banach space
of bounded n-linear mappings on   E   with values in the space   F   is
denoted by L_n(E,F).   The Banach space of bounded n-linear mappings
on   E   with values in the space   F   is denoted by L_n(E,F),   and if
necessary is canonically identified with $L(E,L_{n-1}(E,F))$ by the
linear isometry   σ   defined by $σ(T)(x)(y_1,y_2,\ldots,y_{n-1})$ =
$T(x,y_1,\ldots,y_{n-1})$.

If   G   is an open subset of   E,   and   f:G → F   is a continuous
function then   f   is said to be differentiable at a point   x ∈ G   if
there is a linear operator T_x:E → F   such that

$$\lim_{\substack{\|h\|\to 0 \\ h\neq 0}} \frac{f(x+h) - f(x) - T_x(h)}{\|h\|} = 0.$$

The linear operator T_x is called the derivative of   f   at   x   and
is denoted by   Df(x).   If   Df(x)   exists for all   x ∈ G,   then   f
is said to be once differentiable in   G.   If the mapping   Df:G →
L(E,F)   is continuous,   f   is said to be of class c^1 in   G.   If
$D^{n-1}f$:G → $L(E,L_{n-2}(E,F))$ is defined and is of class c^1 in   G   so
that D^nd:G → $L(E,L_{n-1}(E,F))$ is a continuous mapping, then   f   is
said to be of class c^n.   It is known [4], that if D^nf(x)   exists
then, for all   n ≥ 2, D^nf(x)   is symmetric, i.e., $D^n f(x)(y_1,\ldots,y_n)$
is independent of rearrangements of $(y_1,\ldots,y_n)$.

If the norm functional of   E   is once differentiable in
E ~ {0,1},   then it is known that it is of class c^1,   Cudia [4].
In this paper   E   is said to be smooth if the norm is differentiable
in   E ~ {0}.   If   x ∈ E, $\|x\|$ = 1,   and if   g´(x)   is the derivative
of the norm at   x,   then it is known that   g´(x)   is the support
functional of   U   at   x. E_x is used to denote the closed subspace

$\{y \mid g'(x)(y) = 0\}$. If the norm is K times differentiable $K \geq 1$, at x, then the derivative of the norm at x is denoted by $g^K(x)$, and is a K-linear symmetric real-valued function. E is K-smooth if the norm is K-times differentiable at all points $x \in E$, $x \neq 0$, and the K-th differential g^K is a function on $E \sim \{0\} \to L_K(E,R)$.

A Banach space E is said to be K-boundedly smooth if it is K-smooth and $\sup_{x \in S} \|g^K(x)\| < \infty$.

A Banach space E is said to be superreflexive if it is isomorphic to a uniformly smooth Banach space (as defined in Day [5]). For an account of superreflexive spaces see James [9], Day [6], and Enflo [8].

Section 2. Classification of Reflexive Spaces

A Banach space E is 0-reflexive if it is reflexive. It is 1-reflexive if it is superreflexive. It is K-reflexive, $K \geq 2$, if it is isomorphic to a Banach space F, where F is K-boundedly smooth. It is w-reflexive if it is isomorphic to a Banach space F which is K-boundedly smooth for every natural number K.

Lemma 1. Let the norm in E be K-times differentiable at x. Then it is K-times differentiable at λx, $\lambda \neq 0$, and D

$$D^K \|\lambda x\| = \frac{|\lambda|}{\lambda^K} D^K \|x\|.$$

The lemma is a consequence of the absolute homogenity of the norm functional.

Lemma 2. If $f:[-1,1] \to E$ is a K-times differentiable function, and if $\|D^K f(t)\| \leq M_K$, $\|f(t)\| \leq M_0$ for all $t \in [-1,1]$ then for all n, $1 \leq n \leq K - 1$, $\sup_{t \in [-1,1]} \|D^n f(t)\| \leq M_n$ where M_n depends only on n, M_0 and M_K.

The lemma follows from Taylor's theorem and induction. In this connection, see page 188 [7].

Theorem 1. If E is a p-boundedly smooth, Banach space then it is K-boundedly smooth for $1 \leq K \leq p$.

Proof. If $K = 1$, then since $\|g'(x)\| = 1$ for all $x \in S$, E is 1-boundedly smooth. Let $K \geq 2$. Let $x_0 \in S$, and $\|h\| \leq 1/2$. If G is the ring $\{x \mid 1/2 \leq \|x\| \leq 3/2, \; x \in E\}$, then there is a constant M such that $\|g^p(x)\| \leq M$ for all $x \in G$ as a consequence of Lemma 1 together with the fact that E is p-boundedly smooth. Thus for $t \in [-1,1] = J$, $\|g^p(x_0 + th)\| \leq M$. Further if $f(t) = \|x_0 + th\|$, $t \in J$, then $f^K(t) = g^K(x_0 + th)(h,h,\ldots,h)$ if $1 \leq K \leq p$. Since $\|x\| \leq 3/2$ for $x \in G$, it follws from Lemma 2, that $|f^K(0)| = |g^K(x_0)\underset{K}{(h,h,\ldots,h)}| \leq C$ is independent of the choice of $x_0 \in S$. Since for each x_0, $g^K(x_0)$ is a symmetric K-linear mapping, it follows from the preceding that $\|g^K(x_0)\| \leq c_1$ for some constant c_1 independent of x_0. Thus E is K-boundedly smooth for $1 \leq K \leq p$.

The following theorem is known, Leonard and Sundaresan [10] and is stated here for convenience of reference.

Theorem 2. If E is 2-boundedly smooth E is uniformly smooth. Thus E is 1-reflexive.

The following theorem summarizes the preceding observations.

Theorem 3. If E is p-reflexive, then E is K-reflexive for $1 \leq K \leq p$.

The following example illustrates the importance of the preceding classification.

Examples. (1) If $1 < p < \infty$, the Banach spaces ℓ_p, $L_p(\mu)$ (where μ is a positive measure) are w-reflexive if p is an even integer. If $p = 2n + 1$, these spaces are 2n-reflexive. If p is not an integer these spaces are [p]-reflexive where [p] is the largest integer less than p.

(2) If $p = 2n + 1$, $1 < p < \infty$, ℓ_p, $L_p(\mu)$ (where μ is a positive measure not supported by finitely many disjoint atoms) are not K-reflexive for any integer $K > 2n$. If p is not an integer

ℓ_p, $L_p(\mu)$, μ as in the preceeding sentence, are not K-reflexive for any $K > [p]$.

For (1) see Sundaresan [12] and Bonic and Frampton [2]. For the first part of (2) see [2]. The proof of the last part of (2) is similar to the first part, and details are in Sundaresan [14].

Before concluding this section it is noted that the Banach space C_o is isomorphic with a Banach space $(E, \| \ \|)$ such that the $\| \ \|$ is of class C^∞, [2]. However since C_o is not reflexive, $D^2 \| \ \|$ is not bounded in S_E.

Section 3. <u>Curvature of the Unit Sphere in a Banach Space</u>

This section is particularly devoted to 2-smooth Banach spaces as the second order derivative is significant in the study of the curvature of the sphere. There are several concepts of curvature of a curve in an abstract metric space, Blumenthal [1]. Here the usual concept of curvature of a smooth curve defined in terms of the first two derivatives is used.

If E is a 2-smooth Banach space and $x \in S$, then the curvature of x in the direction of y, $\|y\| = 1$, $x \neq y \neq -x$, is by definition $g^2(x)(x,y)$. One can verify that the usual curvature of the curve obtained by the section of S with the two-dimensional space spanned by $\{x,y\}$ at x is $g^2(x)(x,y)$. Since $g^2(x)$ is a positive form, the curvature is non-negative. The functions ρ_i, $i = 1,2$ on S into non-negative reals are defined by

$$\rho_1(x) = \sup_{\|y\|-1} g^2(x)(y,y) = \|g^2(x)\|,$$

and

$$\rho_2(x) = \inf_{\|y\|=1, y \in E_x} g^2(x)(y,y).$$

Before proceeding to the discussion of curvature, functions ρ_1 and ρ_2 in terms of second order derivatives of the norm, certain useful results are proved. These results attempt to generalize Mazur's theorem to the effect that if

$$\lim_{t \to 0} \frac{\|x + ty\| - \|x\|}{t} = F(y)$$

exists for all $y \in E$, for some $x \neq 0$, then F is a continuous linear functional, and is the tangent functional of the unit cell at $x | \|x\|$.

Lemma 3. Let E be a normed linear space, and let $x \in S$, the unit sphere of E; further let the norm of E be once directionally differentiable, and $G(x)$ be the directional derivative of the norm at x. Then the following are equivalent.

(i) $\displaystyle \lim_{t \to 0} \frac{\|x + t\xi\| - \|x\| - tG(x)\xi}{t^2} = 0$

(ii) $\displaystyle \lim_{t \to 0} \frac{\|x + t\xi\| - 1}{t^2} = 0.$

Proof. It is clear that (i) implies (ii). To prove that (ii) implies (i), note that $G^{-1}(x)(0) = E_x$ is a maximal subspace of E, every vector $\xi \in E$ admits a unique representation $\xi = \lambda x + h$ where $h \in E_x$. Thus if $\|x\| = 1$,

$$\frac{\|x + t\xi\| - 1 - tG(x)}{t^2} = \frac{\|x + \frac{th}{1 + t\lambda}\| - 1}{t^2/(1 + t\lambda)^2} \cdot \frac{1}{(1 + t\lambda)} .$$

Proceeding to the limit as $t \to 0$, the proof is completed.

Proposition 1. Let E be a two-dimensional normed linear space which is 1-smooth. In order that the norm in E is twice differentiable at x, $\|x\| = 1$, it is necessary and sufficient that $\displaystyle \lim_{t \to 0} \frac{\|x + th\| - 1}{t^2}$ exists for each $h \in E_x$.

First a symmetric bilinear form T_x on $E \times E$ is constructed so that

$$(1) \quad \|x + t\xi\| = \|x\| + tg'(x)\xi + t^2 T_x(\xi, \xi) + 0_x(t)$$

where $\dfrac{O_x(t)}{t^2} \to O$ as $t \to O$. Let T_x' be defined such that

$$T_x'(x,h) = T_x'(h,x) = T_x'(x,x) = O \quad \text{and} \quad T_x'(h,h) = \lim_{t \to O} \dfrac{\|x + th\| - 1}{t^2}.$$

These values of T_x' define a unique bilinear form on T_x on $E \times E$. It is verified that if $\xi = \lambda x + \mu h$, then

$$\lim_{t \to O} \dfrac{\|x + t\xi\| - tG(x)\xi - 1}{t^2} = \mu^2 \lim_{t \to O} \dfrac{\|x + th\| - 1}{t^2} = \mu^2 T_x'(h,h)$$

$$= T_x'(\xi, \xi).$$

Further from the definition of T_x it follows that

$$T_x(\xi, \xi) = \mu^2 T_x'(h,h).$$

From these it is verified that T_x is a symmetric bilinear operator and verifies the condition (1). Since E is two-dimensional and x,h are linearly independent it is verified that the norm is twice differentiable with $g^2(x) = T_x$ The next corollary is a generalization of Mazur's theorem to the second order derivative of the norm, and is a consequence of the preceding proposition.

Corollary 1. The norm of the Banach space E is twice differentiable at $x \neq O$, if and only if

$$\lim_{t \to O} \dfrac{\|x + t(y + z)\| - \|x + t(y - z)\|}{t^2} = F(y,z)$$

exists as a continuous bilinear form on $E \times E$.

From Proposition 1, it follows that the two-dimensional sectional curvature of S, the unit sphere of a Banach space E, exists at a point x if and only if $\lim\limits_{t \to O} \dfrac{\|x + ty\| - \|x\|}{t^2}$ exists for all $y \in E$, and this limit is the curvature of S at x in the direction of y. The following proposition is a consequence of Proposition 1, and Theorem 4 [12].

Proposition 2. If a Banach space E is 2-smooth at a point $x \in S$, and if the two-dimensional sectional curvatures at x in the directions $y \in E_x$ is uniformly bounded below by a number $c > 0$, then E is isomorphic to a Hilbert space. In particular it is w-smooth.

From the definition of a 2-reflexive space it follows that if a Banach space E is 2-reflexive then there is a convex body X in E such that the two-dimensional sectional curvatures of the boundary of X is uniformly bounded above.

The relations between curvature, K-reflexivity of a Banach space E with a Schauder basis, and basis are in Sundaresan [14].

REFERENCES

[1] Blumenthal, L.M., _Theory and Application of Distance Geometry_, Oxford Claredon Press, 1953.

[2] Bonic, R. and J. Frampton, 'Smooth functions on a Banach manifold', J. of Math. and Mech. 15 (1966), 877-898.

[3] Cartan, H., _Differential Calculus_, Herman, Paris, 1971.

[4] Cudia, D., 'The geometry of Banach spaces', Trans. Amer. Math. Soc. 110 (1964), 284-314.

[5] Day, M.M., _Normed Linear Spaces_, 3rd Ed., Ergeb. der Math. Vol. 21, Springer-Verlag, 1973.

[6] Day, M.M., 'Mimicry in Banach spaces', these Proceedings.

[7] Dieudonne, J., _Foundations of Modern Analysis_, Academic Press, New York, 1960.

[8] Enflo, P., 'Banach spaces which can be given an equivalent uniformly convex norm', Israel J. Math 13 (1972), 281-288.

[9] James, R.C., Superreflexive Banach Spaces, Canadian J. Math. Vol. XXIV (1972), 896-904.

[10] Leonard, I.E. and K. Sundaresan, 'Geometry of Lebesgue-Bochner functions spaces - Smoothness', Bull. Amer. Math. Soc. 79 (1973), 546-549.

[11] Leonard, I.E. and K. Sundaresan, 'Geometry of Lebesgue-Bochner function spaces', to appear in Trans. Amer. Math. Soc. (1974).

[12] Sundaresan, K., 'Smooth Banach spaces', Math. Annalen 73 (1967), 191-199.

[13] Sundaresan, K., 'Some geometric properties of the unit cell in spaces C(X,B)', Bulletin of Polish Academy of Sciences XIX (1971), 1007-1012.

[14] Sundaresan, K., 'Approximation in K-reflexive spaces', to appear.

NORM IDENTITIES WHICH CHARACTERIZE
INNER PRODUCT SPACES

John Oman
University of Wisconsin

It is well known that if a nontrivial norm identity holds in a normed linear space then the norm is determined by an inner product. For some norm identities, M.M. Day has shown that if the identity holds only for unit vectors the space must still be an inner product space. This paper gives similar results for vectors restricted to sets other than the unit sphere. For example, it is shown that if there exists $0 < \lambda < 1$ such that the identity $\|\lambda x + (1 - \lambda)y\|^2 + \lambda(1-\lambda)\|^2 = \lambda\|x\|^2 + (1 - \lambda)\|y\|^2$ holds for all vectors x and y in some set K with nonempty interior then the space is an inner product space.

The paper also extends work of D.A. Senechalle and previous work of the author to localize Day's theorem. The result proven here is that if $\sim$ is one of the relations $\geq$ and $\leq$, $\varepsilon > 0$, and $0 < u < 1$ then a real normed linear space X is an inner product space if and only if for $x, y \in X$ with $\|x\| = \|y\| = 1$ and $\|x - y\| < \varepsilon$ there exists $0 < \lambda < 1$ such that

$$u(1 - u)\|\lambda x + (1 - \lambda)y\|^2 + \lambda(1 - \lambda)\|ux - (1 - u)y\|^2 \sim$$

$$(\lambda + u - 2\lambda u)(\lambda u + (1 - \lambda)(1 - u)).$$

The method of proof is the standard one of showing that the unit sphere is an ellipsoid. In order to use this method the paper proves that if S is a plane symmetric closed convex curve then there exist many inscribed and circumscribed ellipses whose contact contains at least two independent points.

NORM IDENTITIES WHICH CHARACTERIZE INNER PRODUCT SPACES

0. Introduction

The best known characterization of inner product spaces among normed linear spaces is by Jordan and von Neuman and states that $\|x + y\|^2 + \|x - y\|^2 = 2\|x\|^2 + 2\|y\|^2$) characterizes inner product spaces. An immediate consequence of this result is that a space is an inner product space if and only if each of its two-dimensional subspaces is an inner product space.

Considerable work has been done proving that other norm identities also characterize inner product spaces. M.M. Day proved that for some identities these results can be improved in three ways:

1. The identity need not hold for all vectors in the space.

2. The identity can be weakened to an inequality.

3. The identity may vary with the choice of vectors as long as the form of the identity is fixed.

Theorem 0. Let $\sim$ be one of the relations $\geq$ and $\leq$. A real normed linear space X is an inner product space if and only if for each pair of vectors $x, y \in X$ with $\|x\| = \|y\| = 1$ there exist real numbers λ and u with $0 < \lambda, u < 1$ such that

$$u(1 - u)\|\lambda x + (1 - \lambda\|^2 + \lambda(1 - \lambda)\|ux - (1 - u)y\|^2 \sim$$
$$[\lambda + u - 2\lambda u][\lambda u + (1 - \lambda)(1 - u)].$$

Most of the characterizations in this paper are similar to the one above because they prove that if an identity holds for some restricted set of vectors then the space is an inner product space. The method of proof is the usual one of showing that for two-dimensional spaces the unit sphere (i.e. the Minkowski gauge curve) must be an ellipse. Section 1 develops some results on inscribed and circumscribed ellipses for centrally symmetric convex curves which are used in the proofs of some of the characterizations.

1. Ellipses

Let S be a plane closed convex curve symmetric about the origin O. An inscribed (circumscribed) ellipse E with center O is said to be an inscribed (circumscribed) n-ellipse if the contact (i.e. E ∩ S) contains at least n independent points. The Loewner inscribed (circumscribed) ellipse is the inscribed (circumscribed) ellipse of maximal (minimal) area and is known to be a 2-ellipse. In this section we find other 2-ellipses and determine some of their properties. The procedure used is very similar to that used for the Loewner ellipses.

If x is a point on the ellipse E with center O let x* denote one of the points of E on the conjugate diameter to the diameter at x. The shorter arc from x to x* is denoted xx*. Let A(x,x*) denote the union of the arcs xx* and -(xx*) (i.e. the reflection in the origin of xx*.) A subset k of E is said to be spread with respect to x if it is not contained in the interior of A(x,x*) or in the interior of A(x,-x*).

Theorem 1, like a corresponding theorem for Loewner ellipses by Behrend, does not directly give inscribed 2-ellipses but does construct inscribed ellipses whose contact is spread.

Theorem 1. Let S be a plane closed convex curve symmetric about the origin O, and let w ≠ O be a point inside of S. Then there exists an ellipse E(w) whose area is maximal among all ellipses with center O, containing w, and inscribed in S. Moreover, the contact of E(w) is spread with respect to w.

Proof. The existence of E(w) follows from usual arguments and clearly the contact is nonempty.

To show the second part of the conclusion we show that if the contact of an inscribed ellipse E with center O and through w is contained in the interior of A(w,w*) then the area of E is not maximal. Without loss of generality we may assume the equation

of E is $x^2 + y^2 = 1$, that $w = (\sqrt{2}/2, \sqrt{2}/2)$ and $w^* = (-\sqrt{2}/2,$
$\sqrt{2}/2)$. Since the contact is contained in the interior of $ww^* \cup -$
$(-ww^*)$ there exists $\delta > 0$ such that if $(x,y) \in S$ and $|y/x| \leq 1$
then $x^2 + y^2 \geq 1 + \delta$. Consider the ellipse E' whose equation is
$(1/(1+\delta))x^2 + ((1 + 2\delta)/(1 + \delta))y^2 = 1$. This ellipse is inscribed
in S, contains w, and has area greater than that of E. Q.E.D.

It should be noted that unlike the Loewner inscribed ellipse
there need not be a unique inscribed ellipse containing w of maximal area. Thus $E(w)$ denotes any one of these maximal ellipses.

Clearly $E(w)$ is an inscribed 2-ellipse unless the only points
of contact are w^* and $-w^*$. Theorem 2 shows that there are an infinite number of inscribed 2-ellipses if S is not an ellipse.

Theorem 2. Let $w \in S$ and w not be in the contact of the
Loewner inscribed ellipse. Then there exists $0 \leq r_0 < 1$ so that
$E(rw)$ is an inscribed 2-ellipse for $r_0 < r < 1$.

Proof. If the theorem is not true then there exists a sequence
$r_i \to 1^-$ such that $E(r_i w)$ are 1-ellipses. Some convergent subsequence of the contact points must converge to a point $v \in S$
where v has a support line parallel to line Ow. Thus a subsequence of the ellipses $E(r_i w)$ converges to an ellipse E whose
contact contains v and w. Since v and w are the endpoints of
conjugate diameters, a result of Behrend implies E is the Loewner
inscribed ellipse contrary to the choice of w. Q.E.D.

Theorem 3 proves that most of the points of S are in the
contact of some inscribed 2-ellipse.

Theorem 3. Let $I(s)$ denote the points of S which are in
the contact of some inscribed 2-ellipse. Then $I(s)$ is dense in
S.

Proof. Let $w \in S \setminus I(s)$ and $r_i \to 1^-$. By theorem 2 we may
assume $E(r_i w)$ are 2-ellipses. Unless all points of contact of
$E(r_i w)$ converge to w, a subsequence of the ellipses $E(r_i w)$ would

converge to a 2-ellipse E with w in its contact. Thus w is the limit of a sequence of contact points of 2-ellipses. Q.E.D.

Theorem 4 is one of the theorems used to prove the localization of Day's theorem found in section 3. It shows that there exist 2-ellipses with close contact points.

Theorem 4. If S is not an ellipse then for any $\epsilon > 0$ there exists at least two inscribed 2-ellipses with the property that the contact of these ellipse contain points p and q with $|p - q| <$ ϵ.

Proof. If $S \smallsetminus I(s) \neq \emptyset$ the result follows from the proof of theorem 3.

Thus assume $S = I(s)$. Let E be any inscribed 2-ellipse and p and q be any points in the contact of E such that the interior of the arc pq contains no points of S. If w is any point on the arc of S between p and q then w is in the contact of some 2-ellipse E'. The second point of contact must also be on the arc of S between p and q since E and E' can intersect in at most four points. Repeat this argument as often as necessary to obtain an ellipse E' with the desired property. Replacing E by E' in the above argument a second ellipse E" with the desired property is obtained. Q.E.D.

The next four theorems show that similar results hold for circumscribed ellipses. The proofs are much the same except that some details require additional care and some theorems require the additional hypothesis that S be strictly convex.

Theorem 5. Let S be a plane closed convex curve symmetric about O and let w be a point outside of O. There exists an ellipse E'(w) whose area is minimal among all ellipses with center O, containing w, and circumscribed about S. Moreover, the contact of E'(w) is spread with respect of w.

Proof. Again the existence follows from the usual arguments

and we prove the second part of the conclusion by showing that if the contact of an ellipse E is contained in the interior of A(w,w*) then the area is not minimal.

Let the equation of E be $x^2 + y^2 = 1$, $w = (1,0)$ and $w* = (0,1)$. There exists $r > 0$ such that the arc from $(r, (1 - r^2)^{1/2})$ to $w* = (0,1)$ contains no points of S, and there exists $0 < \delta < 1 - ((1 - r)/(1 + r))^{1/2}$ such that if $(x,y) \in S$ and $y/x \leq 0$ or $y/x \geq (1 - r^2)^{1/2}/r$ then $x^2 + y^2 \leq 1 - \delta$.

The ellipse $x^2 + Bxy + Cy^2 = 1$ where $B = 2((1 - r)/(1 + r))^{1/2}(1 - 1/(1 - \delta)^2)$ and $C = (2/(1 - \delta)^2 - 1) + 1/r)/(1 + 1/r)$ circumscribes S, contains w, and has smaller area than E. Q.E.D.

Theorem 6 is stated without proof since the proof is essentially the same as that for theorem 2.

Theorem 6. Let $w \in S$ but not be in the contact of the Loewner circumscribed ellipse. Then there exists $r_0 > 1$ so that $E'(rw)$ is a circumscribed 2-ellipse for

$$1 < r < r_0.$$

Theorem 7 and 8 correspond to theorems 3 and 4 respectively. These theorems require the additional hypothesis that S be strictly convex. The proof of theorem 7 is sketched and the proof of theorem 8 omitted.

Theorem 7. Let C(S) denote the points of S which are in the contact of some circumscribed 2-ellipse. If S is strictly convex then C(S) is dense in S.

Proof. The only difference from theorem 3 is that the contact points of the $E'(r_i w)$ may converge to points distinct from w but the limit ellipse may be degenerate. This, however, implies that S contains a segment contrary to the hypothesis. Q.E.D.

Theorem 8. If S is strictly convex and not an ellipse then

for any $\epsilon > 0$ there exist at least two circumscribed 2-ellipses with points p and q in the contact of the ellipse such that $|p - q| < \epsilon$

2. Generalizations of Day's Theorem

The following theorem is a simple generalization of Day's theorem. Basically it states that Day's theorem is valid for more general identities and for sets K other than the unit sphere as long as the set K contains vectors in all directions. The result is stated without proof since only minor modifications in Day's elegant proof are needed. The sums involved may be infinite as well as finite if everything converges.

Theorem 9. Let $\sim$ be one of the relations $\geq$ and $\leq$. A real normed linear space X is an inner product space if and only if there exists a set K with the property that for each $x \in X$ there exists $\alpha \in R$ such that $\alpha x \in K$ and such that for each $x, y \in K$ there exist real numbers a_ν, b_ν, and c_ν (which may depend on x and y) such that

1. $a_\nu > 0$
2. $\sum_\nu a_\nu b_\nu c_\nu = 0$
3. $b_\nu c_\nu \neq 0$
4. $\sum_\nu a_\nu \|b_\nu x + c_\nu y\|^2 \sim \sum_\nu a_\nu (b_\nu^2 \|x\|^2 + c_\nu^2 \|y\|^2)$

The condition that $a_\nu > 0$ restricts rather strongly the type of identities that can be used in this theorem. D.A. Senechalle in [7] and S.O. Carlsson in [2] have worked with identities where $a_\nu \neq 0$ but the identity must hold for more vectors in the space.

Theorem 9 can also be extended to identities involving more than two vectors.

Theorem 10. Let $\sim$ be one of the relations $\geq$ and $\leq$. A real normed linear space X is an inner product space if and only if there exists an integer n and a set K with the property that for each $x \in X$ there exists $\alpha \in R$ such that $\alpha x \in K$ and such that for each $x_1, \ldots, x_n \in K$ there exist real numbers a_ν and $b_{\nu i}$ such that

1. $a_\nu > 0$

2. $\sum_\nu a_\nu b_{\nu i} b_{\nu j} = 0 \quad i \neq j$

3. for each ν, $b_{\nu 1} \neq 0$ and $\sum_{i=2}^{n} b_{\nu i} \neq 0$

4. $\sum_\nu a_\nu \| \sum_{i=1}^{n} b_{\nu i} x_i \|^2 \sim \sum_\nu a_\nu \sum_{i=1}^{n} b_{\nu i}^2 \|x_i\|^2$

Proof. Let $x_2 = \ldots = x_n$ and the theorem reduces to theorem 9. Q.E.D.

The following corollary to theorem 10 is a generalization theorem 2 in [8] by K. Sundaresan.

Corollary 11. Let X be a normed linear space, $x' = \sum_{i=1}^{n} x_i / n$ if $x_i \in X$, and $\sim$ be one of the relations $\geq$ and $\leq$. Then X is an inner product space if and only if there exists an $n \geq 2$ such that

$$\|x'\|^2 + (1/n) \sum_{i=1}^{n} \|x_i - x'\|^2 \sim 1$$

for $x_i \in X$ with $\|x_i\| = 1$.

3. Localization of Day's Theorem

Theorem 12 extends the results of Senechalle in [6] and Oman in [5] by localizing theorem 9. The identity is assumed to hold only for unit vectors close together. Since theorem 9 is in a sense the best possible the identity in theorem 12 is slightly less general than the identity in theorem 9 and the set K in theorem 9 is the unit sphere in theorem 11.

Theorem 12. Let $\epsilon > 0$; $0 < u < 1$, and $\sim$ be one of the relations $\geq$ and $\leq$. A real normed linear space X is an inner product space if and only if for $x, y \in X$ with $\|x\| = \|y\| = 1$ and $\|x - y\| < \epsilon$ there exist real numbers λ, d_1, d_2, a_ν, b_ν, and

c_ν such that

1. $0 < \lambda < 1$

2. $d_1, d_2 > 0, a_\nu \geq 0$

3. $\sum_\nu a_\nu b_\nu c_\nu = d_2 u(1 - u) - d_1 \lambda(1 - \lambda)$

4. $d_1 \|\lambda x + (1 - \lambda)y\|^2 + d_2 \|ux - (1 - u)y\|^2 + \sum_\nu a_\nu \|b_\nu x + c_\nu y\|^2$

$\sim d_1(\lambda^2 + (1 - \lambda)^2) + d_2(u^2 + (1 - u)^2) + \sum_\nu a_\nu(b_\nu^2 + c_\nu^2).$

Proof. If X is an inner product space the identity clearly holds. Also, it suffices to prove the theorem when ·X is two-dimensional.

Consider the case where $\sim$ is $\geq$. Let E be an inscribed ellipse in the unit sphere S with x and y in the contact of E such that $\|x - y\| < \epsilon$. (The distance here is Minkowski distance while the distance in theorem 4 was Euclidean distance, but this is no problem since the two metrics are equivalent.) If $|\cdot|$ is the norm determined by E then $|z| \geq \|z\|$ for all $z \in X$ and $|\cdot|$ is determined by an inner product. Thus

$d_1 \|\lambda x + (1 - \lambda)y^2 + d_2 \|ux - (1 - u)y\|^2 + \sum_\nu a_\nu \|b_\nu x + c_\nu y\|^2 \geq$

$d_1(\lambda^2 + (1 - \lambda)^2 + d_2(u^2 + (1 - u)^2) + \sum_\nu a_\nu(b_\nu^2 + c_\nu^2) =$

$d_1 |\lambda x + (1 - \lambda)y|^2 + d_2 |ux - (1 - u)y|^2 + \sum_\nu a_\nu |b_\nu x + c_\nu y|^2 \geq$

$d_1 \|\lambda x + (1 - \lambda)y\|^2 + d_2 \|ux - (1 - u)y\|^2 + \sum_\nu a_\nu \|b_\nu x + c_\nu y\|^2.$

Hence, equality must hold throughout so

$|\lambda x + (1 - \lambda)y| = \|\lambda x + (1 - \lambda)y\|$

$|ux - (1 - u)y| = \|ux - (1 - u)y\|$

and $|b_\nu x + c_\nu y| = \|b_\nu x + c_\nu y\|$. By repeating this argument and using the fact that $\lambda x + (1 - \lambda)y$ lies "between" x and y it follows that the whole arc xy is in the contact of E. If $E \neq S$

the contact must consist of arcs and points such that the "distance" between them is greater than ϵ.

Several cases depending on the value of u must be considered separately.

Suppose $0 < u < 1/2$. Let x and y be unit vectors such that y is an endpoint of one of the above arcs, x lies on this arc, and let $x \to y$. Then $w = -(ux - (1 - u)y)/\|ux - (1 - u)y\|$ is in the contact of E and $w \to y$ through points not on the arc. But this contradicts the fact that no other point of contact can be within ϵ of the arc.

Suppose $u = 1/2$. Let x, y and w be as above. This time $w \to z$ where z is parallel to a support line to the unit sphere at y. But this is known to imply that E is the Loewner inscribed ellipse. From theorem 4 there is a second ellipse with the same properties as E so it must also be the Loewner ellipse. Since the Loewner ellipse is unique this also is a contradication.

Suppose $1/2 < u < 1$. Interchange x and y in the case where $0 < u < 1/2$ and a similar contradication is reached.

Thus E = S and X is an inner product space.

Now consider the case where $\sim$ is $\leq$. An argument similar to the one above holds once it has been determined that X must be strictly convex.

If X is not strictly convex then vectors x and y can be chosen so that:

1. $\|x\| = \|y\| = 1$ and $\|x - y\| < \epsilon$

2. $\|\lambda x + (1 - \lambda)y\| = 1$ for $0 < \lambda < 1$

3. $\|ux - (1 - u)y\| > 1 - 2u$.

Under these conditions $\|bx + cy\|^2 \geq (b + c)^2$ for all real numbers b and c. Thus

$$d_1\|\lambda x + (1 - \lambda)y\|^2 + d_2\|ux - (1 - u)y\|^2 + \sum_\nu a_\nu\|b_\nu x + c_\nu y\|^2 >$$

$$d_1 + d_2(1 - 2u)^2 + \sum_\nu a_\nu (b_\nu + c_\nu)^2 =$$

$$d_1(\lambda^2 + (1 - \lambda)^2) + d_2(u^2 + (1 - u)^2) + \sum_\nu a_\nu (b_\nu^2 + c_\nu^2)$$

contrary to the hypothesis. Q.E.D.

It is interesting to note that in the above proof $\sum_\nu a_\nu \|b_\nu x + c_\nu y\|^2$ is not really used but is simply carried along. Also Senechalle, using different techniques, has worked with $u = 1/2$ and in the above proof $u = 1/2$ had to be handled separately from $u \neq 1/2$. Finally, in theorem 11 the values of u is fixed while the other real numbers may depend on x and y. The following example shows that theorem 11 is false if u is also allowed to depend on x and y.

Example 13. Consider the two-dimensional Minkowski space whose unit sphere (i.e. Minkowski gauge curve) is given by the following equations.

$$4x^2 + 4/3y^2 = 1 \quad \text{if} \quad 3|x| \leq |y|$$
$$|x| + |y| = 1 \quad \text{if} \quad 1/3|x| \leq |y| \leq 3|x|$$
$$4/3x^2 + 4y^2 = 1 \quad \text{if} \quad |y| \leq 1/3|x|$$

In this space if $\|z\| = \|w\| = 1$ and $\|z - w\| < 1/4$ then there exist $0 < \lambda, u < 1$ such that

$$u(1 - u)\|\lambda z + (1 - \lambda)w\|^2 + \lambda(1 - \lambda)\|uz - (1 - u)w\|^2 =$$
$$(\lambda + u - 2\lambda u)(\lambda u + (1 - \lambda)(1 - u)).$$

If both z and w are on the same elliptical arc or on the same line segment the existence of λ and u is trivial. If one is on an elliptical arc and the other is on an adjacent line segment the existence of λ and u follows from a continuity argument.

4. Characterizations Using Two-Dimensional Subspaces

Theorem 12 like most characterizations of inner product spaces is based on the fact that a space is an inner product space if each

two-dimensional subspace is an inner product space. The hypothesis
of the characterizations in sections 5 and 6 are such that it cannot
be shown that every two-dimensional subspace is an inner product
space. Theorem 15 proves the unsurprising result that it suffices
to have many two-dimensional subspaces which are inner product
spaces.

Lemma 14. A real normed linear space X is an inner product
space if and only if there exist a hypersubspace H and a vector
x not in H such that H is an inner product space and every two-
dimensional subspace containing x is an inner product space.

Proof. If X is an inner product space clearly x and H
exist.

Suppose x and H exist. Let H′ be any closed hypersubspace such that x/‖x‖ + H′ is a supporting hyperplane to the unit
sphere at x/‖x‖. Since every two dimensional subspace through x
is an inner product space it follows that $\|rx + h'\|^2 = \|rx\|^2 +$
$\|h'\|^2$ for real numbers r and all vectors h and H′.

Next it is shown that H′ is an inner product space by proving
that the parallelogram law holds in H′. Let h′, g′ ε H′ and
chose real numbers r and s so that rx + h′, sx + g′ ε H. Since
the parallelogram law holds in H it follows that

$$\| (rx + h') + (sx + g') \|^2 + \| (rx + h') - (sx + g') \|^2 =$$

$$2\|rx + h'\|^2 + 2\|sx + g'\|^2.$$

Applying the Pythagorean relation it follows that

$$(r + s)^2\|x\|^2 + \|h' + g'\|^2 + (r - s)^2\|x\|^2 + \|h' - g'\|^2 =$$

$$2r^2\|x\|^2 + 2\|h'\|^2 + 2s^2\|x\|^2 + 2\|g'\|^2$$

or that

$$\|h' + g'\|^2 + \|h' - g'\|^2 = 2\|h'\|^2 + 2\|g'\|^2.$$

Finally, it is shown that the parallelogram law holds in X. Let $ax + h'$, $bx + g' \in X$ where $a, b \in R$ and $h', g' \in H'$.

$$\|(ax + h') + (bx + g')\|^2 + \|(ax + h') - (bx + g')\|^2 =$$

$$(a + b)\|x\|^2 + \|h' + g'\|^2 + (a - b)^2\|x\|^2 + \|h' - g'\|^2 =$$

$$2(a^2 + b^2)\|x\|^2 + 2\|h'\|^2 + 2\|g'\|^2 =$$

$$2\|ax + h'\|^2 + 2\|bx + g'\|^2. \quad \text{Q.E.D.}$$

<u>Theorem 15</u>. A real normed linear space X is in inner product space if and only if there exists a basis $\{y, x_\alpha\}$ for X such that every two-dimensional subspace which contains an x_α is an inner product space.

<u>Proof</u>. Using lemma 14 as the inductive step the result follows for all finite dimensional spaces. Since any two-dimensional subspace is contained in some finite dimensional subspace generated by the basis vectors the result follows for infinite dimensional spaces. Q.E.D.

5. <u>Identities on a Set K</u>

Theorem 9 shows that the unit sphere can be replaced by any set K which contains vectors in all directions. Are there other sets which can replace the unit sphere? Theorems 16 and 17 answer this question in the affirmative. Again, the identity satisfied must be more specific since the set of vectors is more restricted.

<u>Theorem 16</u>. A normed linear space X is an inner product if and only if there exists a subset K of X with nonempty interior with the property that for $x, y \in K$ there exists $0 < \lambda < 1$ such that

$$\|\lambda x + (1 - \lambda)y\|^2 + \lambda(1 - \lambda)\|x - y\|^2 = \lambda\|x\|^2 + (1 - \lambda)\|y\|^2.$$

<u>Proof</u>. If X is an inner product space let $K = X$.

Assume such a set K exists. Since any subset of K has the same property as K, it suffices to prove the result for balls $B = B(z, \varepsilon) = \{x: \|x - z\| < \varepsilon\}$. To begin assume X is two dimensional.

Let $y \in B$ such that $| \|y\| - \|z\| | < \|y - z\| < \|y\| + \|z\|$. Let E be the ellipse with center 0 and which contains $y/\|y\|$, $z/\|z\|$, and $(y - z)/\|y - z\|$ and let $|\cdot|$ be the norm determined by E. Let $C(z,y) = \{az + by : a, b \geq 0\}$ denote the closed convex cone determined by z and y, and let S again denote the unit sphere. If $C(z,y) \cap S \cap E \neq C(z,y) \cap E$ then there exist $y' = \lambda y + (1 - \lambda)z$ and $z' = \lambda'y + (1 - \lambda')z$ such that $C(y',z') \cap S \cap E = \{y'/\|y'\|, z'/\|z'\|\}$.

Since B is convex $y', z' \in B$ so there exists $0 < \lambda'' < 1$ so that

$$\|\lambda''z' + (1 - \lambda'')y'\|^2 + \lambda''(1 - \lambda'')\|z' - y'\|^2 =$$

$$\lambda''\|z'\|^2 + (1 - \lambda'')\|y'\|^2. \quad \text{But} \quad \|y'\| = |y'|,$$

$$\|z'\| = |z'|, \quad \text{and} \quad \|z' - y'\| = |z' - y'| \quad \text{so}$$

$$\|\lambda''z' + (1 - \lambda'')y'\| = |\lambda''z' + (1 - \lambda'')y'|$$

contrary to the choice of y' and z'. Hence $C(y,z) \cap S \cap E = C(y,z) \cap E$ (i.e. this partition of S is an elliptical arc).

Let $w = (1/2)(z + y)$ and $x \in X$. There exists $\delta > 0$ such that $w + \delta x \in C(y,z) \cap S(z, \varepsilon)$. By hypothesis there exists $0 < \lambda < 1$ such that

$$\lambda(1 - \lambda)\|\delta x\|^2 = \lambda\|w\|^2 + (1 - \lambda)\|w + \delta x\|^2 - \|\lambda w + (1 - \lambda)(w + \delta x)\|^2$$

$$= \lambda|w|^2 + (1 - \lambda)|w + \delta x|^2 - |\lambda w + (1 - \lambda)(w + \delta x)|^2$$

$$= \lambda(1 - \lambda)|\delta x|^2.$$

Hence $S = E$ and X is an inner product space.

If the dimension of X is greater than two, choose a basis $\{x_\alpha\}$ for X such that $\{x_\alpha\} \subseteq K$. The result now follows from theorem 15. Q.E.D.

Theorem 17 specializes the set K to a convex cone with nonempty interior. The identity in theorem 17 is more general than the identity in theorem 16. Since the two proofs are quite similar the proof of theorem 17 is omitted.

<u>Theorem 17</u>. A normed linear space X is an inner product space if and only if there exist a convex cone K with nonempty interior and a fixed $0 < u < 1$ such that for $x, y \in K$ there exists $0 < \lambda < 1$ such that

$$u(1 - u)\|\lambda x + (1 - \lambda)y\|^2 + \lambda(1 - \lambda)\|ux - (1 - u)y\|^2 =$$
$$\lambda u(\lambda + 2\lambda u)\|x\|^2 + (1 - \lambda)(1 - u)(\lambda + u - 2\lambda u)\|y\|^2.$$

The identities in theorems 16 and 17 are the weakest for which the theorems are valid. To check this consider the space in example 13. If $K = B((1/2, 1/2), .01)$ and u is very close to either 0 or 1 then all $x, y \in K$ and all $0 < \lambda < 1$ it follows that

$$u(1 - u)\|\lambda x + (1 - \lambda)y\|^2 + \lambda(1 - \lambda)\|ux - (1 - u)y\|^2 =$$
$$\lambda u(\lambda + u - 2\lambda u)\|x\|^2 + (1 - \lambda)(1 - u)(\lambda + u - 2\lambda u)\|y\|^2.$$

Thus in theorem 16 the vector $x - y$ cannot be replaced by $ux - (1 - u)y$.

If $K = C((1/3, 2/3), (2/3, 1/3))$ then for $x, y \in K$ and for all $0 < \lambda, u < 1$ it follows that

$$u(1 - u)\|\lambda x + (1 - \lambda)y\|^2 + \lambda(1 - \lambda)\|ux - (1 - u)y\|^2 \geq$$
$$\lambda u(\lambda + u - 2\lambda u)\|x\|^2 + (1 - \|)(1 - u)(\lambda + u - 2\lambda u)\|y\|^2.$$

Thus in theorems 16 and 17 "$=$" cannot be replaced by "$\geq$".

Finally if $K = C((-1/4,3/4),(1/4,3/4))$ then for $x,y \in K$ and for all $0 < \lambda, u < 1$ it follows that $u(1 - u)\|\lambda x + (1 - \lambda)y\|^2 + \lambda(1 - \lambda)\|ux - (1 - u)y\|^2 \leq \lambda u(\lambda + u - 2\lambda u)\|x\|^2 + (1 - \lambda)(1 - u)(\lambda + u - 2\lambda u)\|y\|^2$. Thus in theorems 16 and 17 "=" cannot be replaced by "$\leq$".

6. Restriction of Only One Vector

The previous sections have discussed characterizations of the type where an identity holds for all vectors x and y subject to some restrictions. In theorem 18 only one vector is restricted.

Theorem 18. A real normed linear space X is an inner product space if and only if there exists a set K and real numbers $0 < \lambda, u < 1$ such that Lin $\{K\}$ is a hypersubspace of X and for $x \in K$ and $y \in X$ then

$$\lambda(1 - \lambda)\|ux - (1 - u)y\|^2 + u(1 - u)\|\lambda x + (1 - \lambda)y\|^2 \leq$$
$$\lambda u(u + \lambda - 2u\lambda)\|x\|^2 + (1 - \lambda)(1 - u)(u + \lambda - 2u\lambda)\|y\|^2.$$

Proof. If X is an inner product space let $K = X$.

Suppose K exists. By theorem 15 it sufficies to prove the theorem for two-dimensional spaces. Let E be any inscribed 2-ellipse in the unit sphere S, and let $|\cdot|$ be the norm determined by E. For two-dimensional spaces it can be assumed that K consists of one vector X. It is now shown that $x/\|x\|$ is in the contact of E.

Suppose w and v are in the contact of E. If $x = aw + bv$ then let $y = (ua/(1 - u))w - (\lambda b/(1 - \lambda))v$. Then $\lambda x + (1 - \lambda)y$ is a scalar multiple of w and $ux - (1 - u)$ is a scalar multiple of v so

$$u(1 - u)\|\lambda x + (1 - \lambda)y\|^2 + \lambda(1 - \lambda)\|ux - (1 - u)y\|^2 =$$
$$u(1 - u)|\lambda x + (1 - \lambda)y|^2 + \lambda(1 - \lambda)|ux - (1 - u)y|^2 =$$
$$\lambda u(\lambda + u - 2u\lambda)|x|^2 + (1 - \lambda)(1 - u)(\lambda + u - 2u\lambda)|y|^2 \geq$$

$$\lambda u(\lambda + u - 2u\lambda) \|x\|^2 + (1 - \lambda)(1 - u)(\lambda + u - 2u\lambda) \|y\|^2.$$

Thus equality must hold and $\|x\| = |x|$ so $x/\|x\|$ is in the contact of E.

But the above argument holds for any inscribed 2-ellipse E. Since only one inscribed 2-ellipse can contain a given point of S in its contact this implies S is an ellipse. Q.E.D.

It is easy to construct examples where Lin {K} does not contain a hypersubspace so that theorem 18 is no longer valid. Moreover, if X is the ℓ_1 plane and $x = (0,1)$ then for all $y \in X$

$$\|x + y\|^2 + \|x - y\|^2 \geq 2\|x\|^2 + 2\|y\|^2$$

so "$\leq$" in theorem 18 cannot be replaced by "$\geq$". Theorem 18 is one of the few examples of a characterization in which the identity can be weakened to an inequality in only one direction.

REFERENCES

[1]   F. Behrend, Uber die Kleninste umschriebene und die grosste
      einschriebene Ellipse eines konrexen Bereichs, Math. Ann. 115
      (1938), 379-411.

[2]   S.O. Carlsson, Orthogonality in normed linear spaces, Ark.
      Mat. 4 (1962), 297-318.

[3]   M.M. Day, On criteria of Kasahara and Blumenthal for inner
      product spaces, Proc. Amer. Math. Soc. 10 (1959), 92-100.

[4]   _____, Normed Linear Spaces, Springer Verlag (1958).

[5]   J. Oman, Characterizations of inner product spaces, Ph.D.
      Dissertation, Michigan State Univ., East Lansing, Michigan,
      1969.

[6]   D.A. Senechalle, Euclidean and non-Euclidean norms in a plane,
      Ill. J. Math. 15 (1971), 281-289.

[7]   _____, Equations which characterize inner product
      spaces, Proc. A.M.S. 40 (1973), 209-214.

[8]   K. Sundaresan, Characterizations of inner product spaces,
      Math. Student 29 (1961), 41-45.

Victor Klee
University of Washington

Introduction

Throughout this note, M denotes a metric space and xy the distance between points of M. A <u>chain from</u> x <u>to</u> y <u>in</u> M is a sequence $(x_o, \ldots, x_k)$ of points of M such that

(1) $k \geq 2$, $x_o = x$, $x_k = y$, $x_{i-1}x_i > 0$ for $1 \leq i \leq k$,

and

(2) $\{x_1, \ldots, x_{k-1}\} \cap \{x_o, x_k\} = \emptyset$.

The chain's <u>ratio-sequence</u> is the sequence

$$\rho(x_o, \ldots, x_k) = (x_o x_1 / \sigma, \ldots, x_{k-1} x_k / \sigma),$$

where $\sigma = \sum_1^k x_{i-1}x_i$. The set of all such ratio-sequences of length k,

$$M_k(x,y) = \{\rho(x_o, \ldots, x_k) : (x_o, \ldots, x_k) \text{ a chain from } x \text{ to } y$$
$$\text{in } M\},$$

is contained in the open $(k-1)$-simplex

$$T_k = \{t = (t_1, \ldots, t_k) \in R^k : \text{all } t_i > 0, \ \sum_i^k t_i = 1\}.$$

Plainly $M_2(x,y) = T_2$ if $x \neq y$ and M is connected, while

(3) $M_k(x,y) = T_k$ for all $k \geq 2$ if $x \neq y$

and M is complete and metrically convex. It follows from a theorem of Schoenberg [7, pp.719-722] (see also Blumenthal [1, pp.73-74]) that (3) holds when M is an arc and hence when M is arcwise connected. The purpose of this note is to

show that (3) continues to hold when M is continuumwise connected
but may fail when M is merely connected. In particular, for each
countable subset U of T_3 there exist a dense connected subset
M of the Euclidean plane R^2 and a dense subset E of M such
that $U \cap M_3(x,y) = \emptyset$ for each pair (x,y) of distinct points of
E. These results answer questions raised by Raymond Freese (motivated in part by his paper [2]) at Michigan State University's 1974
conference on metric geometry, honoring Leonard Blumenthal.

Sufficiency of Continuumwise Connectedness

When $\varphi(0) = x$ and $\varphi(1) = y$ the following is a consequence
of Schoenberg's theorem [7].

Theorem 1. Suppose that $x,y \in M$ with $x \neq y$, the mapping
$\varphi:[0,1] \to M$ is continuous, and $s = (s_1,\ldots,s_k) \in T_k$ with
$x\varphi(0) < s_1(x\varphi(0) + \varphi(0)y)$ and $(1)y < s_k(x\varphi(1) + (1)y)$. Then
there exist numbers a_i with $0 < a_1 < \ldots < a_{k-1} < 1$ such that

$$\rho(x,\varphi(a_1),\ldots,\varphi(a_{k-1}),y) = s.$$

Proof. Let the definition of $\rho(x_o,\ldots,x_k)$ be applied to
every sequence $(x_o,\ldots,x_k)$ for which $\sum_1^k x_{i-1}x_i > 0$; then
$\rho(x_o,\ldots,x_k) \in \bar{T}_k$, the closure of T_k. For each $t = (t_1,\ldots,t_k) \in \bar{T}_k$ let

$$\zeta(t) = \rho(x,\varphi(t_1),\varphi(t_1 + t_2),\ldots,\varphi(\sum_1^{k-1} t_i),y) \in \bar{T}_k,$$

whence the mapping $\zeta:\bar{T}_k \to \bar{T}_k$ is continuous. Note that

$$t_i = 0 \Rightarrow \zeta(t)_i = 0 \quad \text{when} \quad 1 < i < k,$$

while

$$t_i = 0 \Rightarrow \zeta(t)_1 \leq \frac{x\varphi(0)}{x\varphi(0) + \varphi(0)y} < s_1$$

and

$$t_k = 0 \Rightarrow \zeta(t)_k \le \frac{(1)y}{x\ (1)\ +\ (1)y} < s_k.$$

Hence the restriction $\zeta|_{\partial T_k}$ maps the boundary ∂T_k into $\bar{T}_k \sim \{s\}$ and for each $t \in \partial T_k$ the segment $[t, \zeta(t)]$ misses s. But then $\zeta|_{\partial T_k}$ is homotopic in $\bar{T}_k \sim \{s\}$ to the identity mapping ζ on ∂T_k, so if $s \notin \zeta\bar{T}_k$ it follows from Borsuk's homotopy extension theorem [3, p.86] that ζ extends to a continuous mapping $\eta: \bar{T}_k \to \bar{T}_k \sim \{s\}$. The composition of η with radial projection (from s) onto ∂T_k is then a retraction of $\bar{T}_k$ onto ∂T_k, and since there is no such retraction [3, p.40], it follows that $s \in \zeta(\bar{T}_k)$. Let $t \in \bar{T}_k$ with $\zeta(t) = s$ and set $a_j = \sum_1^j t_i$ for $1 \le j \le k - 1$. Then it is evident that $0 < a_1 < \ldots < a_{k-1} < 1$ and

$$\rho(x, \emptyset(a_1), \ldots, (a_{k-1}), y) = s. \qquad \square$$

Theorem 2. **If the distinct points** x **and** y **lie together in a continuum** K **in** M **then** $M_k(x,y) = T_k$ **for all** $k \ge 2$.

Proof. Consider an arbitrary $s = (s_1, \ldots, s_k) \in T_k$ and choose $\delta > 0$ such that

$$(4) \quad 0 < \delta < xy \min(s_1, s_2).$$

By a well-known result [8, p.16] there exist a subcontinuum C of K and points $u, v \in C$ such that

$$(5) \quad xu = \delta = vy.$$

We may assume without loss of generality that K lies in a Banach space B. For each n it follows from C's connectedness that there is a chain $(c_o, \ldots, c_{\ell(n)})$ from u to v in C with $c_{j-1}c_j < 1/n$ for $1 \le j \le \ell(n)$. Now consider the polygon

$$P_n = \bigcup_1^{\ell(n)} [c_{j-1}, c_j] \subseteq B$$

and note that

(6) P_n lies in the $1/n$-neighborhood of the compact set   C.

Plainly there exists a continuous $\varphi:[0,1] \to P_n$ with $\varphi(0) = 0$ and
$\varphi(1) = v$,   so (in view of (4) and (5)) Theorem 1 guarantees the
existence of points $x_1(n),\ldots,x_{k-1}(n)$ of P_n such that

$$\rho(x,x_1(n),\ldots,x_{k-1}(n),y) = s.$$

In view of (6) we may assume there are points $x_i \in C$ such that
$\lim\limits_{n\to\infty} x_i(n) = x_i$ for $1 \le i \le k - 1$,   and then $(x,x_1,\ldots,x_{k-1},y)$
is a chain from   x   to   y   in   K   with

$$\rho(x,x_1,\ldots,x_{k-1},y) = s. \quad \square$$

Insufficiency of Connectedness

Let $\underset{\sim}{C}$ denote the set of all continua   C   in the Euclidean
plane R^2 such that   C   does not lie entirely in the "vertical"
line (line of the form $\{r\} \times R$ for $r \in R$)   and $\underset{\sim}{C}'$ the set of
all $C \in \underset{\sim}{C}$ such that   C   does not lie in the union of fewer than
$2^{\aleph_0}$ lines and circles.   Note that all parabolic hyperbolic, and
noncircular elliptic arcs belong to $\underset{\sim}{C}'$.

Theorem 3.   Suppose that for each $k \ge 3$, $U_k \subset T_k$,
card $U_k < 2^{\aleph_0}$,

$$V_k = \{t \in U_k : t_i/t_j \text{ is irrational for } 1 < i < j < k\}$$

and

$$W_k = \{t \in U_k : t_i/t_j \text{ is irrational for } 1 \le i < j \le k\}.$$

Then there exist sets   D   and   E   and an additive function $\varphi:R \to R$
such that the following six conditions are satisfied when   S   is the
graph

$$G_\varphi = \{(r,\varphi(r)):r \in R\}:$$

(a) S is connected;

(b) $D \subset S$ and $E \subset R^2 \sim S$;

(c) $W_k \cap S_k(x,y) = \emptyset$ for each $k \geq 3$ and each pair (x,y) of distinct points of D;

(d) $V_k \cap (S \cup E)_k(x,y) = \emptyset$ for each $k \geq 3$ and each pair (x,y) of distinct points of E;

(e) $card(S \cap C) = 2^{\aleph_0}$ for each $C \in \underset{\sim}{C}$;

(f) $card(D \cap C) = card(E \cap C) = 2^{\aleph_0}$ for each $C \in \underset{\sim}{C}'$.

The key property (a) of Theorem 3 follows from (e) in conjunction with a theorem of F.B. Jones [4] asserting G_φ is connected for each additive $\varphi:R \to R$ such that G_φ intersects every member of $\underset{\sim}{C}$. Note that every point of R^2 is a condensation point of both D and E. The sets S and $S \cup E$ are both connected and the former, being a rational linear subspace of R^2 is also metrically convex.

Proof of Theorem 3. In preparation for a later use, we define the set

$$\Gamma(p,q,\mu) = \{z \in R^2 : \|z - p\| = \mu\|z - q\|\}$$

for each pair (p,q) of points of R^2 and each $\mu > 0$. If $p = q$ and $\mu \neq 1$ the set is empty, while if $p \neq q$ and $\mu = 1$ the set is the perpendicular bisector of the segment $[p,q]$. The following two remarks are also required.

(7) If $p \neq q$ and $\mu \neq 1$ the set $\Gamma(p,q,\mu)$ is the circle whose center and radii are respectively $(u^2q - p)/(u^2 - 1)$ and $u\|p - q\|/|\mu^2 - 1|$.

(8) If $p \neq q$ and $\Gamma(p,q,\mu) = \Gamma(p,q,\bar{\mu})$ then $q = \bar{q}$ and $u = \bar{u}$.

For (7), note that the equation, $\langle z - p, z - p \rangle = u^2 \langle z - q, z - q \rangle$, is equivalent to

$$\langle z, z \rangle - 2 \left\langle \frac{\mu^2 q - p}{\mu^2 - 1}, z \right\rangle = \frac{\langle p, p \rangle - \mu^2 \langle q, q \rangle}{\mu^2 - 1}$$

and then complete the square on the left. The assertion (8) is obvious if u and $\bar{\mu}$ is 1. For the remaining case, assume $p = 0$ and note that if

$$\frac{u^2 q}{\mu^2 - 1} = \frac{\bar{u}^2 q}{\bar{\mu}^2 - 1} \quad \text{and} \quad \frac{\mu \|q\|}{|\mu^2 - 1|} = \frac{\bar{\mu} \|\bar{q}\|}{|\bar{\mu}^2 - 1|}$$

then

$$\frac{\bar{\mu} \|\bar{q}\|}{\mu \|q\|} = \frac{\mu(\bar{\mu}^2 - 1)}{\bar{\mu}(\mu^2 - 1)} = \frac{\|\bar{q}\|}{\|q\|},$$

where $u = \bar{u}$ and $q = \bar{q}$.

For each subset X of R or R^2 let $\text{lin } X$ denote the rational linear extension of X. The triple (D, S, E) that is desired in Theorem 3 is "constructed" by a transfinite induction based on a procedure for extending useful triples, where a triple (D, S, E) is said to be <u>useful</u> if it satisfies conditions (b) – (d) of Theorem 3 as well as the following four conditions:

(9) S is a rational linear subspace of R^2;

(10) E is rationally linearly independent;

(11) $S \cap \text{lin } E = \{(0,0)\}$;

(12) $\text{card}(S \cup E) < 2^{\aleph_0}$.

Three different types of extension are required. For each triple (D, S, E) of subsets of R^2 and each point z of R^2, define

$$(D, S, E)_1[z] = (D \cup \{z\}, \text{lin}(S \cup \{z\}), E),$$

$$(D, S, E)_2[z] = (D, \text{lin}(S \cup \{z\}), E)$$

and

$$(D,S,E)_3[z] = (D,S,E \cup \{z\}) .$$

When

(13) $z \notin \lin(S,E)$

and the triples (D,S,E) and $(D,S,E)_m[z]$ are both useful, the latter is said to be an <u>extension</u> of the former. The extension process is based on the following two results.

(14)   If (D,S,E) is useful the set $\{z \in R^2 : (D,S,E)_m[z]$
       is not an extension of $(D,S,E)\}$

is contained in the union of $< 2^{\aleph_o}$ lines and circles.

(15)   If (D,S,E) is useful the set $\{z \in R^2 : (D,S,E)_2[z]$
       is not an extension of $(D,S,E)\}$

is of cardinality $< 2^{\aleph_o}$.

To establish (14) and (15) we must consider the various ways in which $(D,S,E)_m[z]$ may fail to be an extension of the useful triple (D,S,E). Since $\card \lin(S \cup E) < 2^{\aleph_o}$, attention may be restricted to points z for which (13) holds, and for such z it is plain that only (c) and (d) can cause trouble. If (c) or (d) fails for $(D,S,E)_m[z]$ there exist $k \geq 3$, $t \in T_k$, and points $x_o, \ldots, x_k$ with $x_o \neq x_k$ such that $\rho(x_o, \ldots, x_k) = t$ -- that is,

(16) $$\frac{x_o x_1}{t_1} = \frac{x_1 x_2}{t_2} = \ldots = \frac{x_{k-1} x_k}{t_k}$$

-- and one of the following conditions is satisfied:

(17) $m = 1$, $t \in W_k$, x_o, $x_k \in D \cup \{z\}$, all x_i's are in
       $\lin(S \cup \{z\})$ but not all are in S;

(18) $m = 3$, $t \in V_k$, x_o, $x_k \in E \cup \{z\}$, $\{x_1, \ldots, x_{k-1}\} \cap \{x_o, x_k\}$
       $= \emptyset$, all x_i's are in $S \cup E \cup \{z\}$ but not all are
       in $S \cup E$;

(19) $m = 2$, $t \in W_k$, x_o, $x_k \in D$, all x_i's are in
       $\text{lin}(S \cup \{z\})$ but not all are in S;

(20) $m = 2$, $t \in V_k$, x_o, $x_k \in D$, all x_i's are in
       $\text{lin}(S \cup \{z\}) \cup E$ but not all are in $S \cup E$.

Now for $0 \leq i \leq k$, let $x_i = s_i + \lambda_i z$ with $s_i \in S \cup E$ and λ_i rational, and then for $1 \leq i \leq k$ let $r_i = s_{i-1} - s_i$ and $\delta_i = \lambda_i - \lambda_{i-1}$. Note that

$$\delta_i \neq 0 \Rightarrow x_{i-1}x_i = |\delta_i| \; \|z - r_i/\delta_i\|$$

and

$$\delta_i = 0 \Rightarrow x_{i-1}x_i = \|r_i\|.$$

Under (17) and (18) there exists a such that $\delta_a \neq 0$. Consider, for an arbitrary $b \neq a$, the equality

(21) $$\frac{x_{a-1}x_a}{t_a} = \frac{x_{b-1}x_b}{t_b}$$

from (16). When $\delta_b = 0$, (21) becomes

$$\|z - \frac{r_a}{\delta_a}\| = \frac{t_a}{t_b} \|r_b\|,$$

which constrains z to lie on one of the $< 2^{\aleph_o}$ circles having center in $\text{lin}(S \cup E)$ and radius equal to the product of the distance between two points of $S \cup E$ and the ratio of two coordinates of a point of V_k. When $\delta_b \neq 0$, (21) becomes

$$\|z - \frac{r_a}{\delta_a}\| = \frac{t_a}{t_b} \|z - \frac{r_b}{\delta_b}\|,$$

which constrains z to lie on one of the $2^{\aleph_o}$ sets $\Gamma(p, q, u)$ for

which p,q ∈ lin(S ∪ E) and μ is the ratio of two coordinates
of a point of V_k. Each such set is empty, a line or a circle
except when $t_a = t_b$ and $r_a/\delta_a = r_b/\delta_b$. But if $t_a = t_b$ it
follows from the hypotheses of Theorem 3 that m = 3 and {a,c} ∩
{1,k} ≠ ∅ and then from (9) - (11) that $r_a/\delta_a \neq r_c/\delta_c$. That
establishes (14) for m ∈ {1,3}, and of course for m = 2 the
result (14) is superseded by (15).

Under (17) or (18), $\lambda_o = 0 = \lambda_k$ but not all λ_i's are 0.
Let a be the smallest i such that $\lambda_i \neq 0$ and c the largest
i such that $\lambda_{i-1} \neq 0$, whence a ≠ c and $\delta_a \neq 0 \neq \delta_c$. Choose
b ∈ {1,...,k} ∼ {a,c}. Then the equality

$$\frac{x_{a-1}x_a}{t_a} = \frac{x_{b-1}x_b}{t_b} = \frac{x_{c-1}x_c}{t_c}$$

constrains z as follows:

(22) If $\delta_b = 0$, z lies on the intersection of the circles
whose centers are r_a/δ_a and r_c/δ_c and radii are
$(t_a/t_b)\|r_b\|$ and $(t_c/t_b)\|r_b\|$ respectively.

(23) If $\delta_b \neq 0$, z lies on the intersection of the sets
$\Gamma(r_a/\delta_a, r_b/\delta_b, t_b/t_a)$ and $\Gamma(r_c/\delta_c, r_b/\delta_b, t_b/t_c)$.

Under (19) the two radii in (22) cannot be equal for their
ratio t_c/t_a is irrational by the hypothesis concerning W_k. The
same is true under (20) unless a = 1 or c = k, and in those
cases it follows (recalling that $x_o \neq x_k$) from (9) - (11) that
the centers are distinct. With the aid of (7) - (8), a similar
analysis applies under (20). When two sets are distinct and each
is a circle or a line, their intersection consists of at most two
points; (15) follows, for we have just seen that the set in (15) is
covered by $< 2^{\aleph_o}$ such intersections.

Now let   H   be a Hamel basis   (maximal rationally linearly
independent subset) of   R   and ψ the first ordinal number whose
cardinal is $2^{\aleph_0}$.   Since

$$\text{card } H = 2^{\aleph_0} = 2^{\aleph_0} \text{ card } \underset{\sim}{C}$$

the sets   H   and $\underset{\sim}{C}$ can both be indexed by the positive ordinals
$< \psi$ in such a way that each member of   H   appears once and each
member of $\underset{\sim}{C}$ appears $2^{\aleph_0}$ times.   (That is, the indexing genera-
tes a one-to-one mapping of the ordinals onto   H   and a $2^{\aleph_0}$-to-one
mapping of the ordinals onto $\underset{\sim}{C}$.)   We are going to construct a
system

$$\underset{\sim}{U} = \{ (D_\zeta, S_\zeta, E_\zeta) : 0 \leq \zeta < \psi \}$$

of useful triples such that

$$(24) \quad (D_o, S_o, E_o) = (\varphi, \{ (0,0) \}, \varphi)$$

and the following conditions (25) - (29) are satisfied for
$0 < \zeta < \psi$,   where

$$D_\zeta^< = \bigcup_{\eta < \zeta} D_\eta, \quad S_\zeta^< = \bigcup_{\eta < \zeta} S_\eta \quad \text{and} \quad E_\zeta^< = \bigcup_{\eta < \zeta} E_\eta :$$

(25)   there is a rational linear subspace L_ζ of   R   and an
additive function $\varphi_\zeta : L_\zeta \to R$ such that $h_\zeta \in L_\zeta$ and
$S_\zeta = \{ (r, \varphi_\zeta (r)) : r \in L_\zeta \}$;

(26) $S_\zeta = \text{lin} (S_\zeta^< \cup S_\zeta^*)$,   where S_ζ^* consists of a point of
$C_\zeta \sim S_\zeta$ and at most one other point;

(27)   when $C_\zeta \notin \underset{\sim}{C}'$, $D_\zeta = D_\zeta^<$ and $E_\zeta = E_\zeta^<$;

(28)   when $C_\zeta \in \underset{\sim}{C}'$, $D_\zeta = D_\zeta^< \cap \{ d_\zeta \}$ with $d_\zeta \in C_\zeta \sim D_\zeta^<$;

(29)   when $C_\zeta \in \underset{\sim}{C}'$, $E_\zeta = E_\zeta^< \cup \{ e_\zeta \}$ with $e_\zeta \in C_\zeta \sim E_\zeta^<$.

Note that by (25) - (26) the function φ_ζ is an extension of φ_η for all $\eta < \zeta$. When the entire system $\underset{\sim}{U}$ has been constructed the triple (D_ψ, S_ψ, E_ψ) with be useful except for failing to satisfy condition (12) and hence with $\varphi = U_{\zeta < \psi} \varphi_\psi$ the triple $(D_\psi, \varphi, E_\psi)$ is the one desired for Theorem 3. Only the construction of $\underset{\sim}{U}$ remains.

The reason for using Theorem 1 rather than appealing directly to Schoenberg's theorem was to insure that condition (2) would be satisfied. If, in defining chains, it were also required that

$$0 < i < j < k \Rightarrow x_i \neq x_j,$$

(3) would still hold when M is arcwise connected but for $k \geq 4$ it is unclear what would happen when M is merely continuumwise connected. It is also unclear whether (3) holds (with either definition or chain) when M is assumed merely to be connected and complete or topologically complete.

In all of our discussion, the case $x = y$ has been ignored. Plainly (3) does not extend to that case, but seemingly little is known about the detailed behavior of $M_k(x,x)$ as x ranges over M. See Milgram [5], [6] for some partial results.

Construction of the system $\underset{\sim}{U}$ is initiated by (24) and completed by transfinite induction. Suppose, with $0 < \tau < \psi$, that the partial system $\{(D_\zeta, S_\zeta, E_\zeta) : \zeta < \tau\}$ has been constructed so as to satisfy (25) - (29). Then $(D_\tau^<, S_\tau^<, E_\tau^<)$ is a useful triple, and the useful triple (D_τ, S_τ, E_τ) is constructed in three steps as described in the three paragraphs below.

If $C_\tau \not\in \underset{\sim}{C}'$, let $E_\tau = E_\tau^<$. If $C_\tau \in \underset{\sim}{C}'$ it follows from (14) that there exists $e_\tau \in C_\tau \sim E_\zeta^<$ such that the triple $(D_\tau^<, S_\tau^<, E_\tau)$ is useful with $E_\tau = E_\tau^< \cup \{e_\tau\}$.

If $h_\tau \in L_\tau^< = U_{\eta < \tau} L_\eta$, let $S_\tau' = S_\tau$. Otherwise use (15) to see there exists $q \in R$ such that the triple $(D_\tau^<, S_\tau', E_\tau)$ is useful with

$$S_\tau' = \lin(S_\tau \cup \{(h_\tau, q)\}).$$

Note that S_τ' is the graph of an additive function φ_τ' with domain

$$L_\tau' = \lin(L_\tau^< \cup \{h_\tau\}).$$

If $C_\tau \not\subseteq \underset{\sim}{C}'$ [resp. $C_\tau \not\subseteq \underset{\sim}{C}'$] it follows from (14) [resp. (15)] that there exists $(r,s) \in C \sim S_\tau'$ such that $r \not\in L_\tau'$ and the triple (D_τ, S_τ, E_τ) is useful with

$$S_\tau = \lin(S_\tau' \cup \{(r,s)\})$$

and

$$D_\tau = D_\tau^< \cup \{(r,s)\} \quad [\text{resp. } D_\tau = D_\tau^<].$$

At last the proof of Theorem 3 is complete. $\square$

Now for each class $\underset{\sim}{M}$ of metric spaces, let $\underset{\sim}{M}_k$ denote the set of all $t \in T_k$ such that $t \in M_k(x,y)$ whenever $M \in \underset{\sim}{M}$ and (x,y) is a pair of distinct points of M. When $\underset{\sim}{M}$ is the class of all connected spaces, the results of this note imply $\underset{\sim}{M}_2 = T_2$ $\underset{\sim}{M}_3 = \emptyset$ but they do not completely determine $\underset{\sim}{M}_k$ for $k \geq 4$. In particular, for $k \geq 4$ they leave open the possibility that $(1/k, \ldots, 1/k) \in \underset{\sim}{M}_k$ -- that is, each pair (x,y) of distinct points points of a connected space can be joined by a k-lattice in the sense of [1], [7]. When $\underset{\sim}{M}$ is the class of all metrically convex connected spaces, $\underset{\sim}{M}_k$ is undetermined for all $k \geq 3$, though it follows from Theorem 3 that for each $t \in M_k$ there are two coordinates of t whose quotient is rational.

REFERENCES

[1] L.M. Blumenthal, Theory and Applications of Distance Geometry,
     Oxford University Press, 1953.

[2] R.W. Freese, A convexity property, Pac. J. Math. 18 (1966),
     237-241.

[3] W. Hurewicz and H. Wallman, Dimension Theory, Princeton Uni-
     versity Press, 1941.

[4] F.B. Jones, Connected and disconnected sets and the functional
     equation, f(x) + f(y) = f(x + y), Bull. Amer. Math. Soc. 48
     (1942), 115-120.

[5] A.N. Milgram, Some topologically invariant metric properties,
     Proc. Nat. Acad. Sci. U.S.A. 29 (1943), 193-195.

[6] A.N. Milgram, Some metric topological invariants, Reports of
     a Mathematical Colloquium (2) 5-6 (1944), 25-35.

[7] I.J. Schoenberg, On metric arcs of vanishing Menger curvature,
     Annals of Math. 41 (1940), 715-726.

[8] G.T. Whyburn, Analytic Topology, Amer. Math. Soc. Coll. Pub.
     vol. 28, 1942.

<u>POLYGONS</u>

Branko Grünbaum*
University of Washington

Dedicated With Respect and Affection to Leonard M. Blumenthal

1. Introduction

Polygons have been studied since the dawn of mathematics; in more recent times they appear most often as tools or intermediate stages in various proofs - definition of arc-length, Jordan curve theorem, Cauchy's theorem about integrals of analytic functions - to mention just a few. Naturally there are also many investigations that deal with polygons themselves and for their own sake - but there is in all the literature no usable systematic exposition of the theory of polygons. True, a number of books (the best-known one being Brückner [1900]) purport to deal with "polygons and polyhedra" - but it very quickly becomes obvious that the actual interest of the authors is not in polygons, and that there is very little information about them past the most obvious facts. The last brief attempt at a somewhat systematic treatment was made more than half a century ago by Steinitz [1916, Chapter 1].

In connection with a course on combinatorial geometry I gave last year, I started collecting material on the theory of polygons. Looking through the literature I became convinced that - contrary to widespread opinion - the theory of polygons is an interesting and active field. Many of its problems are very attractive, and due to their conceptual simplicity can serve as accessible (but not trivial) analogues and counterparts of problems in other branches of geometry. Unfortunately, those advantages of polygons are offset by the difficulty in locating the relevant papers, and the absence of even a rough guide to the recent literature on polygons.

* Research supported by National Science Foundation Grant GP-42450.

Although I hope to be able to provide someday such a guide, the following pages have a much more modest aim. We shall discuss, in detail, just two questions from the theory of polygons, the classification of planar polygons, and some regularity properties of skew polygons. For both topics we attempted to provide a survey of known results, bibliographic references, and a few open problems. While there is little causal connection between the two parts of this paper, I hope that each is appealing in its own right, and show the vitality of the field.

$$* \qquad * \qquad *$$

In order to simplify the following discussion we begin with some definitions and notational agreements.

For an integer $n \geq 3$, we define an n-<u>gon</u> in the Euclidean d-space E^d as the figure $P = [x_1, x_2, \ldots, x_n]$ formed by n points $x_1, \ldots, x_n$ in E^d and the n segments $[x_i, x_{i+1}]$, $i = 1, 2, \ldots, n-1$, and $[x_n, x_1]$. The points x_i are called <u>vertices</u> of the n-gon, the segments, its <u>sides</u> or <u>edges</u>. For small values of n traditional names <u>triangle</u>, <u>quadrangle</u>, <u>pentagon</u>, <u>hexagon</u>, <u>heptagon</u>, etc. are customary. If the value of n is not important the term <u>polygon</u> is frequently used. Polygons are understood in a cyclic order, so that $[x_1, x_2, \ldots, x_n]$ is the same as $[x_2, \ldots, x_n, x_1]$ and all other polygons obtained by cyclic permutations of the original vertices. The polygon $[x_n, \ldots, x_2, x_1]$ may or may not be considered as different from $[x_1, x_2, \ldots, x_n]$; depending on the point of view chosen we speak about <u>oriented</u> or <u>unoriented</u> polygons. We shall be concerned with unoriented polygons only. To avoid trivialities we shall also assume that $x_i \neq x_{i+1}$ for $i = 1, 2, \ldots, n$; here and in the sequel we shall understand subscripts to be taken mod n, so that $x_{n+1} = x_1$.

The above definition of polygons is usually credited to Poinsot [1810] and Möbius [1865]; actually, it was given already by Albert Girard [1626] and by A.L.F. Meister [1769] - but neither of those two authors seems to have had much influence on later developments.

The notion of polygons just defined is so general that very little can be done without imposing further restrictions. In the sequel we shall do that, trying to strike a reasonable balance between generality and convenience of exposition.

2. Classification of Planar Polygons

We shall say that a polygon P in the plane E^2 is an ordinary polygon provided no point of E^2 belongs to more than two edges. Since the edges are closed segments which include the vertices of P that are their endpoints, this means that in an ordinary polygon P

(i)   no three edges have a common (interior) point;

(ii)   no two edges have a segment of positive length in common;

(iii)   no vertex belongs to the interior of an edge;

(iv)   no two vertices coincide.

In Figure 1 we show examples of hexagons that are not ordinary.

If x_j is a vertex of a polygon P we shall say that x_j is proper provided the edges $[x_{j-1}, x_j]$ and $[x_j, x_{j+1}]$ are not collinear. The polygon P is proper provided all its vertices are proper.

The hexagons in Figure 2 are not proper.

The first topic we shall discuss is the classification of ordinary and proper n-gons in the plane. In other words, how many different kinds of n-gons are there for $n = 3,4,5,\cdots$?

While it is rather plausible that any reasonable classification scheme should assign all triangles to the same type, some more or less arbitrary decisions have to be made concerning the general definition. What appears as a natural classification to one person may be felt as unreasonable by another. Because of that I shall not dwell on the many different schemes that were proposed during the second half of the 19th century, mainly since I find them to be strange and infertile crosses between the aim of a classification and the authors' preoccupation to "correctly" define the various angles, or angle-sums, or areas.

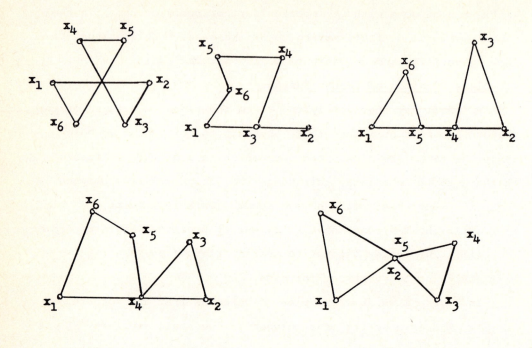

Figure 1

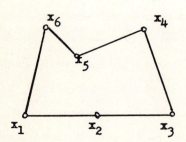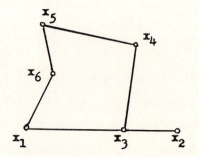

Figure 2

A classification which appears to me as rather natural and
mathematically interesting is the following "isotopy" scheme proposed
by E. Steinitz [1916]:

Orindary and proper n-gons $P_0 = [x_1', \ldots, x_n']$ and $P_1 = [x_1'', \ldots, x_n'']$ in the plane are said to have the same _form_ provided there exists
a family $P(t) = [x_1(t), \ldots, x_n(t)]$, for $0 \leq t \leq 1$, of ordinary and
proper n-gons such that $P(0) = P_0$, while $P(1)$ coincides with P_1
or with a mirror-image of P_1, and for each $j = 1, 2, \ldots, n$ the
point $x_j(t)$ is a continuous function of t in the closed interval
$0 \leq t \leq 1$.

With a little patience it is possible to find that there are 3
different forms of quadrangles and 11 diffferent forms of pentagons.
In Figure 3 we reproduce from Steinitz [1916, Section 5] (see also
Pfeiffer [1937], Du Val [1971]) examples of all the different forms of
quadrangles and of pentagons. Steinitz [1916, p.10] states that the
number of distinct forms of hexagons is 70, an assertion repeated by
Pfeiffer [1937] and Du Val [1971]. However, this claim is not correct since we have:

Theorem 1. There are at least 72 distinct forms of hexagons.

For a proof of Theorem 1 it is enough to consider the 72 hexagons
shown in Figure 4. Since they are easily seen to have distinct _pat-
terns_ (see below), it follows that their forms are distinct.

Theorem 1 may probably be strengthened as follows:

Conjecture 1. There are exactly 72 distinct forms of hexagons.

Nothing seems to be known about the number of distinct forms of
n-gons for $n \geq 7$.

Steinitz's classification by form, unfortunately, does not seem
to have inspired many investigations. In the almost 60 years that
have elapsed, the only mention I could find of it is in the articles
on "Polygons" by Pfeiffer [1937] and Du Val [1971] in two different
editions of "Encyclopedia Britannica": One reason for this neglect

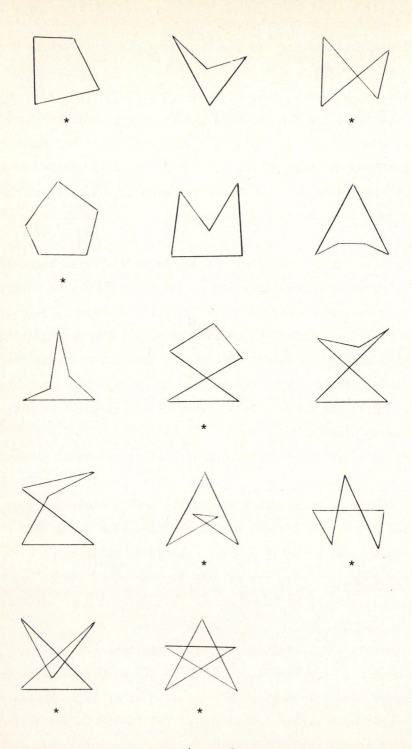

Figure 3

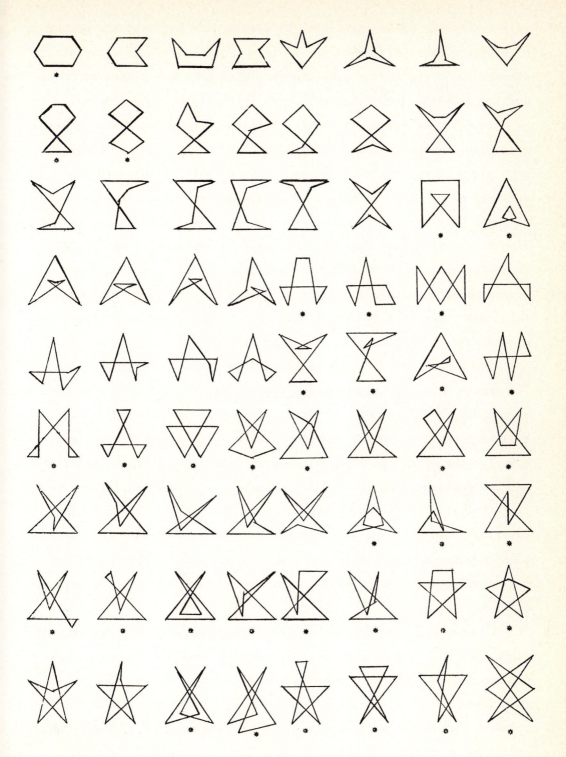

Figure 4

is probably the great difficulty in determining not only the number of distinct forms of n-gons for a given n, but even to ascertain whether two given n-gons have the same form or not. I know of no algorithm or method to decide even that trivially-seeming problem.

In order to arrive at classifications of polygons that are easier to handle, we shall associate with each ordinary n-gon P a <u>planar map</u> M(P) in the following manner: The nodes of M(P) are the vertices of P and the points of selfintersection of P; the edges of M(P) are the parts of the edges of P between the nodes of M(P) (see Figure 5). Thus the nodes of M(P) are either of valence 2 (corresponding to each of the n vertices of P) or of valance 4 (corresponding to the s points of selfintersection of P). Therefore M(P) has n + 2s edges, and s + 2 faces (or countries), one of which is the unbounded exterior of P.

If P and P′ are polygons such that M(P) and M(P′) are isomorphic maps in the <u>unoriented plane</u> (that is, the unbounded faces correspond to each other, and mirror images are not distinguished) we shall say that P and P′ have the same <u>selfintersection-pattern</u>. Clearly, the differentiation of polygons according to their selfintersection-pattern is an equivalence relation, coarser then the equivalence relation of having the same form. Using Figure 3 and 4 it is easily checked that there are at least 2,6,36 different selfintersection-patterns for polygons of 4,5, or 6 sides. Representatives of the different selfintersection-patterns are marked by asterisks in Figures 3 and 4.

On comparing the notions of form and selfintersection-pattern it is obvious that the latter does not distinguish between angles that are smaller than π and those larger than π, while the former does. This leads to the idea of modifying the map M(P) associated with an ordinary proper n-gon P so as to include information regarding which of the two angles at a vertex is less than π. One way of

M(P) :

P :

M'(P) :

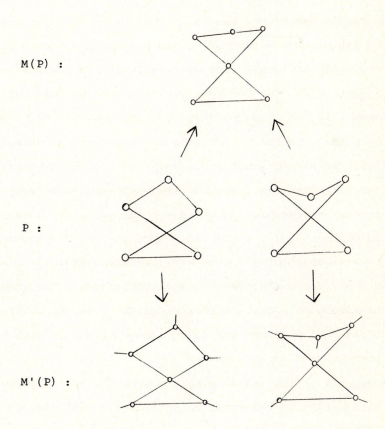

Figure 5

accomplishing this is by attaching a "short" edge leading from each vertex of P into the angle greater than π (see Figure 5). We may denote the planar map obtained in this fashion from P by M'(P), and declare that polygons P_1 and P_2 have the same pattern if $M'(P_1)$ is a planar map isomorphic to $M'(P_2)$.

The equivalence relation defined on polygons by the property of having the same pattern is finer than the equivalence by selfintersection-pattern, but it is coarser (possibly non-strictly) than the equivalence by form. It has the advantage over form to be algorithmically decidable, and we venture the following:

Conjecture 2. Two polygons with the same pattern have the same form.

The classification of polygons by their pattern goes back essentially to Girard [1626], but seems to have been completely forgotten for centuries. Girard [1626] appears to be very rare; a somewhat more accessible, detailed account of Girard's ideas may be found in Günther [1876, pp.17-21], where the contributions of other early investigators of polygons are also discussed.

We shall prove (Theorem 2 below) Conjecture 2 for the special class of simple polygons. But first we shall define this very important class of polygons and discuss some of the results known about them.

A polygon P is called simple provided the only points of the plane that belong to two edges of P are the vertices of P. Hence each simple polygon is topologically equivalent to a simple circuit (that is, it is a topological image of a circle). Equivalently, we may say that an ordinary polygon is simple if and only if it has no selfintersections. One of the fundamental results on simple polygons is the polygonal version of the famous Jordan curve theorem; the case of triangles is important in the axiomatic treatment of Euclidean geometry and is known as the Pasch axiom.

We recall that a polygonal path (also called "open polygon") with endpoints x,y is a set of segments $[x_0, x_1]$, $[x_1, x_2], \ldots,$ $[x_{n-1}, x_n]$ with $x = x_0$ and $y = x_n$. A set X is <u>polygonally connected</u> provided for each pair of points $x, y \in X$ there exists a polygonal path with endpoints x, y that is completely contained in X. A set $X \subset E^2$ is <u>simply connected</u> provided for each topological disc $D \subset E^2$ such that bd $D \subset X$ we have $D \in X$. For sake of completeness, and in order to provide the context for the following results, we shall briefly survey some of the properties of simple polygons.

<u>Proposition 1</u>. (Jordan) If P is a simple polygon in E^2 then the complement of P in E^2 is the union of two disjoint, polygonally connected sets $O(P)$ and $I(P)$, the <u>outside</u> and the <u>inside</u> of P, such that any polygonal path with one endpoint in $O(P)$ and another in $I(P)$ meets P. Moreover, $I(P)$ is simply connected.

Reasonably simple proofs of the polygonal version of the Jordan curve theorem may be found in Aleksandrov [1950, Chapter 1] and Jackson [1968]; for other proofs see Hahn [1908], Lennes [1911], Feigl [1924], Courant-Robbins [1941], Knopp [1945], Lamotke [1958].

Among the consequences of Proposition 1 we mention the following two (compare Lemmas 7 and 8 of Jackson [1968]):

<u>Proposition 2</u>. If the endpoints of a simple polygonal path S belong to a simple polygon P and if all points of S except the endpoints belong to $I(P)$, then the complement of S in $I(P)$ consists of two disjoint regions each of which is the inside of one of the simple polygons formed by S and parts of P between the endpoints of S.

<u>Proposition 3</u>. If P is a simple n-gon with $n \geq 4$ then P has at least one diagonal.

To explain the last assertion we define a $\underline{chord}$ of a simple polygon $P = [x_1,\ldots,x_n]$ as any segment $[x_i,x_j]$ that meets P in x_i and x_j only. A chord $[x_i,x_j]$ is an $\underline{inside\ chord}$ or $\underline{diagonal}$ of P provided its relative interior $]x_i,x_j[$ is contained in $I(P)$, an $\underline{outside\ chord}$ if $]x_i,x_j[\subseteq O(P)$.

A chord $[x_i,x_j]$ of P is called minimal provided $|i - j| = 2$; a minimal chord $[x_i,x_{i+2}]$ is said to $\underline{bridge}$ the vertex x_{i+1}. If $[x_i,x_{i+2}]$ is a minimal chord of P then clearly $I([x_i,x_{i+1},x_{i+2}]) \cap P = \emptyset$. From Proposition 2 and 3 it easily follows by induction:

$\underline{Proposition\ 4}$. If $n \geq 4$, each simple n-gon P has at least two vertices that are bridged by minimal inside chords.

Proposition 4 is very useful in many considerations and it – or facts equivalent to it have been established very often in different contexts. A very simple proof may be derived from the observation (Feigl [1924, Satz 26']) that if a vertex x_i of P is not the endpoint of any diagonal of P then x_i is bridged by a minimal chord of P.

The fact that the inside and the outside of each simple polygon P are well determined may be used to classify the vertices of P according to their "convexity character". Let x_j be a proper vertex of P, so that $[x_{j-1},x_j]$ and $[x_j,x_{j+1}]$ are not collinear. Let $y' = \lambda x_{j-1} + (1 - \lambda)x_j$ and $y'' = \mu x_{j+1} + (1 - \mu)x_j$; it is easily checked that for all sufficiently small $\lambda > 0$ and $\mu > 0$ the open segments $]y',y''[$ are either all contained in $I(P)$, or all in $O(P)$. In the former case we say that x_j is a $\underline{convex}$ vertex, in the latter x_j is a $\underline{concave}$ vertex of P.

A simple polygon $P = [x_1,\ldots,x_n]$ is called $\underline{convex}$ provided $[x_i,x_j]$ is a diagonal of P whenever $|i - j| \geq 2$. The appropriateness of this terminology is shown by the following result (extensions of which to more general sets are well known theorems of the type "local convexity implies convexity"):

Proposition 5. If all vertices of a simple polygon P are convex then P is convex.

Proof. If $z \in]x_i,x_j[$ and $z \in O(P)$, let $x',x'' \in P$ be such that $z \in [x',x'']$ but $]x',x''[\subseteq O(P)$. Then x',x'' are adjacent vertices of a polygon P^* the other vertices and edges of which (besides $[x',x'']$) form an arc of P, such that $I(P^*) \subseteq O(P)$. If y is a vertex of P^* bridged by a minimal chord of P^* and not belonging to $[x',x'']$, then y is a convex vertex of P^* but since y is also a vertex of P it would have to be a concave vertex of P, contradicting the assumptions. If there is no such z then the theorem could fail only if $]x_i,x_j[\subseteq P \cup I(P)$ but $]x_i,x_j[\cap P \neq \emptyset$. In that case either $]x_i,x_j[\subseteq P$ but that is excluded by the assumptions that $|i - j| \geq 2$ and that all the vertices of P are convex, or else $]x_i,x_j[\cap I(P) \neq \emptyset$. In the latter case let $]x',x''[$ be m maximal open segment of $]x_i,x_j[$ that is contained in $I(P)$; then x',x'' are vertices of P, but since $]x',x''[$ is a proper subset of $]x_i,x_j[$ at least one of x',x'' would have to be concave -- contradicting the assumed convexity of all vertices of P.

It is not hard to derive from Proposition 5 the following strengthened version:

Proposition 6. If all vertices of a simple polygon P are convex and if $x,y \in P$ then either $[x,y] \subseteq P$ or else $]x,y[\subseteq I(P)$.

With each simple (not necessarily proper) n-gon $P = [x_1,...,x_n]$ we may associate the signature $\sigma(P)$ of P, which is the sequence $(\sigma_1,...,\sigma_n)$, where

$$
\sigma_j = \begin{cases} 0 & \text{if } x_j \text{ is not a proper vertex} \\ +1 & \text{if } x_j \text{ is a convex vertex} \\ -1 & \text{if } x_j \text{ is a concave vertex} \end{cases}
$$

of P. Since we deal with unoriented polygons we shall not distinguish between signatures obtained from each other by cyclic permutations or by reversal of order.

For proper simple n-gons P the signature $\sigma(P)$ and the map $M'(P)$ determine each other. Hence Conjecture 2 is settled for simple polygons by the following result:

Theorem 2. Simple proper polygons P and P' satisfy $\sigma(P) = \sigma(P')$ if and only if P and P' have the same form.

Proof. A simple proper polygon P is said to have <u>standard shape</u> provided its convex vertices belong to a circle circumscribed about P, and its concave vertices belong to non-intersecting circular arcs (see example in Figure 6).

Since the validity of Theorem 2 is obvious in case P and P' both have standard shape, for a complete proof of the theorem it is enough to show the following:

(*) Each simple proper n-gon P can be continuously deformed under preservation of signature (that is, under conditions set forth in the definition of form) to a polygon that has standard shape. The proof of (*) will be accomplished by induction on n.

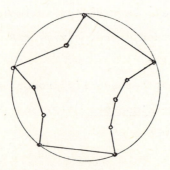

Figure 6

In cases $n = 3$ and $n = 4$ the assertion (*) is obviously true. Assuming its validity for polygons of less than n sides, let the simple proper n-gon P be given. We may clearly assume that P is not convex. Hence P has some outside chord, and therefore also a minimal outside chord $[x_j, x_{j+2}]$. We shall use this minimal outside chord to define a simple $(n - 1)$-gon Q that will be used in the proof. In order to reduce the number of cases we have to consider we shall assume that no three vertices of P are colli-near; this obviously entails no loss of generality since the signa-ture of a simple proper polygon is not affected by sufficiently small perturbations of its vertices. In the first case, the exten-sion of the edge $[x_{j-1}, x_j]$ beyond x_j intersects the edge $[x_{j+1}, x_{j+2}]$ in a point y (see Figure 7; the symmetric situation in which the extension of $[x_{j+2}, x_{j+3}]$ intersects $[x_j, x_{j+1}]$ is

Figure 7

handled by the obvious changes) then Q is obtained by deleting from P the vertices x_j and x_{j+1} and the edges containing them, and introducing instead the vertex y and the edges $[x_{j-1}, y]$ and $[y, x_{j+2}]$. In the other cases the extended edges $[x_{j-1}, x_j]$ and $[x_{j+2}, x_{j+3}]$ do not intersect the triangle with vertices x_j, x_{j+1}, x_{j+2}, the inside of which is in $O(P)$; we delete x_{j+1} and the two edges incident with it and form Q from the remaining part of P by adjoining the edge $[x_j, x_{j+2}]$. The three distinct variants of this situation are shown in Figure 8.

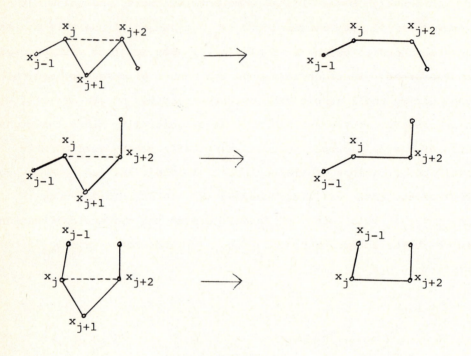

Figure 8

Now, by the inductive hypothesis the simple proper $(n - 1)$-gon Q may be continuously deformed to a standard shape Q^*; by the continuity of the deformation it is obvious that any one edge $[a,b]$ of Q could be replaced by a rhombus $R = [a,c,b,d]$ (see Figure 9) sufficiently thin so that during the deformation from Q

Figure 9

to Q^* the rhombus R is deformed into similar rhombi, and each of those meets the rest of the deformed polygon only at the images of a and b.

Next we observe that P could be continuously deformed into a polygon P' (under preservation of signature) so that in the case represented by Figure 7 we have $[x_{j-1},y] = [a,b]$ and the deformation of the polygonal arc $x_{j-1},x_j,x_{j+1},x_{j+2}$ is to $x_{j-1} = a$, t, $y = b$, x_{j+2}, where t is a suitable point inside the rhombus R; in the cases represented by Figure 8 we take $[a,b] = [x_j,x_{j+2}]$ and move x_{j+1} to a suitable point t inside the rhombus R. After P' has been reached we deform it following the deformation that changed Q to the standard shape Q^*; although the corresponding n-gon P'^* is not in standard shape it is obvious that it can be deformed to a polygon P^* of the same signature and of standard shape, as required.

This completes the proof of Theorem 2.

Among the many open problems related to Conjecture 2 and Theorem 2 the following two appear particularly challenging:

Conjecture 3. If P_1 and P_2 are n-gons with the same selfintersection-pattern then P_1 may be continuously deformed into P_2 through n-gons of the same selfintersection-pattern.

Conjecture 4. If P_1 and P_2 are n-gons with the same pattern and with corresponding edges parallel, then P_1 may be deformed into P_2 through n-gons of the same pattern and with parallel corresponding edges.

In case the n-gons P_1 and P_2 are simple, the validity of Conjecture 4 has been established by Polák [1960]; related material is covered in Polák [1962], [1968, Chapter 8], [1969].

Other open problems involve the classification of not necessarily ordinary polygons, classification by the isomorphism of the

maps induced in the plane by the edges and diagonals of the polygons, and the arrangements of lines generated in the plane by
extending the edges of the polygon. Numerous interesting results
on the last topic are included in the forthcoming Ph.D. thesis of
T. Strommer [1975].

$$* \qquad * \qquad *$$

It is easy to show that each simple polygon has at least 3
convex vertices (this seems to have been first observed by Lionnet
[1873]). Hence the following characterization of signatures of
simple polygons:

Proposition 7. A sequence $(\sigma_1, \sigma_2, \ldots, \sigma_n)$ with each $\sigma_i \in$
$\{+1, 0, -1\}$ is the signature $\sigma(P)$ of a simple n-gon P if and
only if $\sigma_j = +1$ for at least three different subscripts j.

3. Equilateral Polygons

The polygons we shall discuss in this section are not restricted to be planar. Instead in order to avoid considering a field so
wide as to comprise all of knot theory, we shall limit our attention to certain metric properties of polygons that are related to
"regularity" in one sense or another.

Actually, we are concerned with a number of distinct topics,
depending on the kind of regularity we wish to consider. Most of
the notions we shall discuss were discovered several times, the
authors usually not being aware or the relevant work of others.

The first mention of the topic I could find in the mathematical literature is the short note of Auric [1911]. Auric calls an
n-gon P "regular" provided P is equilateral and isogonal, that
is, all edges of P have the same length and the angles between
adjacent edges are all equal. Possibly of greatest interest is
Auric's statement that each "regular" pentagon is planar. Valiron
[1911] noted that skew "regular" n-gons exist for each odd $n > 5$

(their existence for even n ≥ 4 is obvious), and suggested that
the definition of "regular" be modified so as to include the
requirement of equality of the dihedral angles determined at each
edge by the two neighboring edges. Valiron conjectured that if
these dihedral angles are, moreover, assumed to be in the same
sense with respect to an orientation of the polygon, then each
"regular" n-gon will necessarily be planar. (Clearly, both Auric
and Valiron had in mind only polygons in E^3.) An elegant proof
of Valiron's conjecture was kindly communicated to me be Professor
B.L. van der Waerden. Independently Valiron's conjecture has been
formulated, generalized to polygons in spaces of all dimensions,
and proved by Coxeter [1974, pp.4 and 162].

Another independent appearance of the skew "regular" polygons
seems to be the problem posed by Arnold [1957]; I have not seen it
and I am quoting it from Šmakal [1972]. Arnold asked for what n
does there exist in E^3 an equilateral n-gon, isogonal with angle
α. It is not hard to see that Arnold's problem may be answered in
the following manner:

Proposition 8. For each α with $0 < α < α_k = (k - 2)π/k$
there exist equilateral skew n-gons isogonal with angle α for
each even n ≥ k ≥ 4. For each α with $0 < α < α_k$ there exist
equilateral skew n-gons isogonal with angle α for each odd n ≥
max{k,7}. There exist no skew equilateral and isogonal pentagons.

This result, or at least the last part of it, was proved by
Garber-Garvackiĭ-Yarmolenko [1961] (I did not see the paper and I
report on it from the information in Šmakal [1972]). The first two
parts may, in any case, be easily established by using continuity
arguments. As an illustration, a "regular" heptagon isogonal with
angle α = π/2 is indicated in Figure 10.

Still another independent reappearance of the same topic
occurred in van der Waerden [1970], who established that every

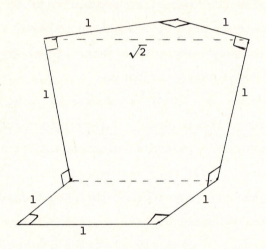

Figure 10

"regular" pentagon is planar. Van der Waerden was motivated to
prove this theorem by interest on part of organic chemists; his
paper provoked a spate of additional proofs (Lüssy-Trost [1970],
Irminger [1970], Dunitz-Waser [1972], van der Waerden [1972],
Šmakal [1972], Kárteszi [1973], Bottema [1973]). Dunitz-Waser
[1972], besides presenting several proofs, note that the result was
established already in Waser [1944], and known to several chemists
at various times. (For the result about pentagons, and for a
slightly weaker form of Proposition 8, see also Waser-Schomaker
[1945].) Kárteszi [1973] also mentions several earlier discoveries
of the planarity of "regular" pentagons; the idea of his proof,
which is the same as that of Bottema [1973], is used in a

generalized form in the proof of Theorem 4 below.

Probably the most elegant proof of the planarity of all "regular" pentagons is the one attributed in van der Waerden [1972] to G. Bol and H.S.M. Coxeter; it is as follows. In each "regular" pentagon P all mutual distances of vertices are determined; therefore each isometry of P is either a motion or a reflection. Let τ be an isometry that cyclically permutes the vertices of P. Then τ^5 is the identity, hence τ is not a reflection but a motion. Since the centroid of P is invariant under τ, the motion τ must be a rotation, and all vertices of P are in one plane orthogonal to the axis of τ. (For elaborations of the idea of this proof see Coxeter [1974, Chapter 1].)

Bottema [1973] completely solved the problem of "regular" pentagons in higher-dimensional spaces by establishing the following result (which may also be derived from the solution of Problems 5 and 6 in Coxeter [1974, p.4]):

Proposition 9. The necessary and sufficient condition for the existence of a skew equilateral pentagon that spans E^4 and is isogonal with angle α is $\pi/5 < \alpha < 3\pi/5$.

If α assumes one of the bounds in Proposition 9 the pentagon becomes planar - either for the regular pentagram, or the regular convex pentagon.

Before reporting on new results that generalize Propositions 8 and 9, we shall survey the investigations that deal with related matters.

As is easily checked, for somewhat elevated values of n there exist both planar and skew n-gons that are "regular" while exhibiting very few features that are intuitively associated with that word. The reason is that the "local" conditions of equilaterality and isogonality impose only very weak global restrictions.

Free from this shortcoming is the following definition, patterned after the one customary in the theory of regular polytopes; in it flag means a pair composed of a vertex and an edge that contains that vertex.

An n-gon P in E^d is called regular provided for each two flags in P there exists an isometry of P onto itself that maps the first flag onto the second.

This definition clearly amounts to saying that each directed edge of P may be mapped onto every other directed edge of P by an isometry of P. An equivalent definition of regular polygons is given in Coxeter [1974, Section 1.5]. Already in Coxeter [1937] a similar definition of regular skew polygons (and polyhedra) appears limited to those without selfintersection.

Motivated by the question of Arnold [1957], a complete characterization of regular n-gons was given by Efremovič-Ilyašenko [1962]. Part of their result may be formulated as follows:

Proposition 10. If P is a regular n-gon that spans E^d then:

(i) If d is odd then $n = 2k$ is even and P is obtained either from a regular n-gon that spans E^{d-1} by moving its vertices perpendicularly by the same length alternately above and under E^{d-1}, or else from two parallel copies of a regular k-gon that spans E^{d-1} the edges of which are replaced by cross-connections.

(ii) If $d = 2p$ is even there exist p pairwise orthogonal 2-dimensional planes $\Pi_1, \ldots, \Pi_p$ such that the projection of P into Π_j parallel to all Π_i with $i \neq j$ is a regular n_j-gon P_j, with n_j a divisor of n and with n equal to the least common multiple of the numbers $n_1, \ldots, n_p$.

Efremovič-Ilyašenko [1962] also completely characterize those sets of planar regular polygons $P_1, \ldots, P_p$ that lead to polygons P that span E^d. For related material see Coxeter [1974, Section 1.7].

The four distinct types of regular 10-gons that span E^3 are indicated in Figure 11.

We shall say that an n-gon $P = [x_1, x_2, \ldots, x_n]$ in E^d is k-equilateral (with parameters c_j) if $|x_i - x_{i+j}| = c_j$ for all i and for $j = 1, 2, \ldots, k$.

Clearly, "2-equilateral" means the same as "equilateral and equiangular". Proposition 9 may be formulated (Bottema [1973]) as asserting that a 2-equilateral pentagon spans E^4 if and only if its parameters c_1 and c_2 satisfy $(\sqrt{5} - 1)/2 < c_1/c_2 < (\sqrt{5} + 1)/2$.

In terms of the notion of k-equilaterality the results of Propositions 8 and 9 may be put in a more general context, as shown by the following conjectures:

Conjecture 5. k-equilateral (2k + 1)-gons span only even-dimensional spaces; each space of dimension $d = 2j \leq 2k$ may be obtained in that way.

Conjecture 6. For each k and each $n > 2k + 1$ there exist k-equilateral n-gons that span E^{2k-1}, but no (k + 1)-equilateral n-gon with odd $n > 2k + 1$ spans E^{2k-1}.

Various parts of these conjectures are known to be true; we shall now present those results. [Since the conjectures were proposed in June 1964, J. Lawrence has established the general case of the first part of Conjecture 5, which is therefore completely solved; Lawrence's paper appears elsewhere in this volume.]

An affirmative solution of the second part of Conjecture 5, essentially contained in Carathéodory [1911], Gale [1963], and

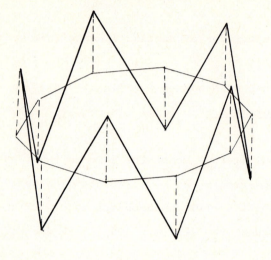

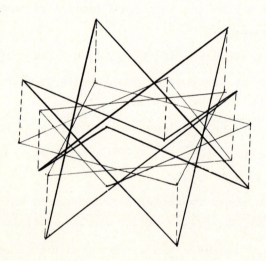

Figure 11 (First Part)

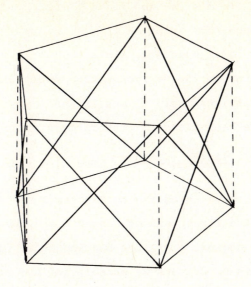

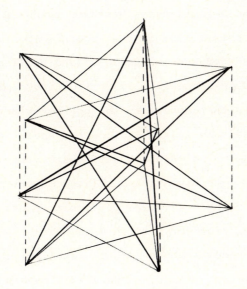

Figure 11 (Second Part)

Coxeter [1974, Chapter 1], is given by the following result:

Theorem 3. If $d = 2k$ and if $n \geq d + 1$, there exists a k-equilateral n-gon that spans E^d.

Proof. Let $x(t) = (\cos t, \sin t, \cos 2t, \sin 2t, \ldots, \cos kt, \sin kt)$ be a point on the "trigonometric moment curve" in E^d. Then the n-gon with vertices $x_i = x(2\pi i/n)$, $i = 0, 1, \ldots, n-1$, is easily seen to be k-equilateral with parameters c_j,

$$c_j^2 = 2k + 1 - \frac{\sin(2k + 1)\pi j/n}{\sin \pi j/n} \quad \text{for} \quad j = 1, 2, \ldots, k,$$

and to span E^d, thus establishing Theorem 3.

Concerning the first part of Conjecture 5, by the results mentioned at the beginning of this section it is valid for $d = 3$. In joint work with O. Bottema we established the following result (the role of which in connection with the proof of Conjecture 5 is by now superseded by the result of Lawrence [1975]):

Theorem 4. No 3-equilateral heptagon spans E^5.

Proof. The central idea in the proof of Theorem 4, kindly suggested to me by Professor O. Bottema (and utilized also in Bottema [1973] and in Kárteszi [1973]) is the use of the Cayley-Menger determinants that appear in the characterization of sets isometrically imbeddable in the Euclidean d-space E^d by conditions involving the mutual distances of their points. A complete treatment of these determinants and related questions is given in Blumenthal [1953, Chapter 4]. The particular facts we need in connection with the proof of Theorem 4 and which, we hope, will be usable in proving results about k-equilateral polygons, are as follows. We denote by $V(P_{2k+1})$ the $(2k)$-dimensional volume of the convex hull of the k-equilateral $(2k + 1)$-gon P_{2k+1}; if P_{2k+1} does not span E^{2k} then $V(P_{2k+1}) = 0$, and we have the alternatives:

(i)   No $2k - 1$ vertices of P_{2k+1} are affinely independent;

then the affine hull of P_{2k+1} is of dimension less than $2k - 2$

and our assertion holds; or else

(ii) E^{2k-2} is the span of some set   T   of $2k - 1$ vertices

of P_{2k+1}.

In the second case, we shall consider the convex hull of each

of the two $(2k)$-pointed sets obtained by adjoining one of the re-

maining vertices of P_{2k+1} to   T.   We shall endeavour to show that

each of those convex hulls has $(2k - 1)$-dimensional content $W = 0$,

hence all vertices of P_{2k+1} span only a $(2k - 2)$-dimensional

space.   Now using the Cayley-Menger determinants it is easy to see

that in the case of a k-equilateral $(2k + 1)$-gon P_{2k+1} with

parameters c_j we have (putting $b_j = c_j^2$):

$$-4^k((2k)!)^2(V(P_{2k+1}))^2 = D_k =$$

$$= \begin{vmatrix} 0 & 1 & 1 & 1 & 1 & \cdots & 1 & 1 & 1 & 1 & \cdots & 1 & 1 \\ 1 & 0 & b_1 & b_2 & b_3 & \cdots & b_{k-1} & b_k & b_k & b_{k-1} & \cdots & b_2 & b_1 \\ 1 & b_1 & 0 & b_1 & b_2 & \cdots & & & \cdots & & \cdots & b_3 & b_2 \\ 1 & b_2 & b_1 & 0 & b_1 & \cdots & & & \cdots & & \cdots & b_4 & b_3 \\ \cdot & \cdot & \cdot & \cdot & \cdot & \cdot & \cdot & \cdot & \cdot & \cdot & \cdot & \cdot & \cdot \\ 1 & b_2 & b_3 & b_4 & b_5 & \cdots & & & \cdots & & \cdots & 0 & b_1 \\ 1 & b_1 & b_2 & b_3 & b_4 & \cdots & & & \cdots & & \cdots & b_1 & 0 \end{vmatrix}$$

and $2^{2k-1}((2k - 1)!)^2 W^2 = \Delta_k$,   where the determinant Δ_k is ob-

tained from D_k by deleting the last row and column.   The proof

is, therefore, reduced to the algebraic assertion that $D_k = 0$ for

some k implies $\Delta_k = 0$;   in case of Theorem 4 we deal with the

case $k = 3$.

Now, for each k it is not hard to show (the author is

indebted to Professor J.S. Frame for kindly pointing out this fact)

that $D_k = -(2k + 1)F_k^2$,   where F_k is a homogeneous polynomial of degree k in $b_1, \ldots, b_k$. More precisely, $F_k = \prod\limits_{j=1}^{k} \sum\limits_{s=1}^{k} (\epsilon^{js} + \epsilon^{-js}) b_s$, where $\epsilon = \exp(2\pi i/(2k+1))$ is a primitive $(2k+1)$st root of unity. In particular, writing for typographical convenience $p = c_1$, $q = c_2$, $r = 0$ we have

$$D_1 = -3p^2 \qquad\qquad , \quad F_1 = p;$$
$$D_2 = -5(p^2 + q^2 - 3pq)^2, \quad F_2 = p^2 + q^2 - 3pq;$$
$$D_3 = -7F_3^2 \qquad\qquad , \quad \text{where}$$
$$F_3 = p^3 + q^3 + r^3 + 3(p^2 q + q^2 r + r^2 p) - 4(pq^2 + qr^2 + rp^2)$$
$$- prq.$$

(The expression for D_2 appears in Bottema [1973] and in Kárteszi [1973], that for D_3 was first communicated to me by O. Bottema.)

For the completion of the proof of Theorem 4 we only need the value of Δ_3. As may be computed with some effort, we have $\Delta_3 = G_3 F_3$ with

$$G_3 = 4(p^2 + q^2 + r^2) - 6(pq + qr + rp),$$

which establishes Theorem 4.

The proof of the general conjecture that $D_k = 0$ implies $\Delta_k = 0$ seems to require more ability and agility in algebraic manipulation than that available to the author.

Much less is known concerning Conjecture 6. In case $n = 2k + 3$ the validity of the second part follows from the result of Lawrence [1975]. Regarding the first part, already Valiron [1911] observed that for each $n \geq 6$ there exist 2-equilateral n-gons that span E^3. It should not be very hard to provide suitable families of examples that will establish the first part of Conjecture 6 in general. However, an experimental approach led to the following observation which appears to deserve a more detailed investigation.

For a fixed choice of the parameters c_1 and c_2, the 2-equilateral heptagons in E^3 are <u>movable</u>, that is, there exists a one-parameter family of them, continuously deformable into each other. However, a closer inspection shows that there actually exist two <u>disjoint</u> families of this kind with the same parameters c_1, c_2, such that heptagons in the same family are continuously deformable into each other but not into any member of the other family. While this assertion is easily verified by experimenting with suitable models, the author sees no easy way to establish it mathematically.

It would appear that many interesting phenomena concerning k-equilateral polygons in various spaces are still awaiting discovery and exploration.

4. <u>Remarks</u>

(1) Proposition 3 is quite old (see some historical data in Simon [1906, pp.166-167]), and it was probably considered self-evident much earlier. The result may be used to establish the possibility of triangulating by $n - 3$ diagonals any simple n-gon. This latter topic was considered recently in Klamkin-Taylor [1970], where it is also shown that there exist, for each $n \geq 3$, n-gons with only $n - 3$ diagonals. It is easy to see that every P with signature $\sigma(P) = (1,1,1,-1,-1,\ldots,-1,-1)$ has this property. However, an n-gon P may have only $n - 3$ diagonals even if its signature is not of this type. It is not known for what signatures there exist polygons with only $n - 3$ diagonals.

(2) It is rather obvious that the proof of Theorem 2 needs only easy modifications in order to adapt it to the situation in which polygons with few selfintersections are considered. Certainly up to three selfintersections can be handled without serious difficulties. Thus a proof of Conjecture 2 for those cases can be

found, but it seems that the resolution in the general case will require radically different methods.

(3) As a consequence of Theorem 2 and Proposition 7, the problem of enumerating the distinct forms of simple n-gons becomes a purely combinatorial question. It differs from the well-known problem of "necklaces" only by the exclusion of sequences containing at most two +1's, which is easily taken into account. For access to the literature on necklaces see Sloane [1973].

(4) We shall call n-$\underline{arc}$ a figure formed in the plane by n distinct points ($\underline{vertices}$) $v_1, v_2, \ldots, v_n = v_0$, and n simple Jordan arcs $A_i = \overset{\frown}{v_{i-1} v_i}$, $i = 1, 2, \ldots, n$, where for $i \neq j$ the intersection $A_i \cap A_j$ is $\underline{either}$ an endpoint of A_i and of A_j, $\underline{or}$ a single point relatively interior to each arc, $\underline{or}$ empty. Thus n-arcs without triple points ($\underline{ordinary}$ n-arcs) are generalizations of ordinary n-gons, in which the rectilinearity assumption has been dropped but the mutual intersections of pairs of arcs are analogous to those of segments.

The definition of $\underline{selfintersection\text{-}pattern}$ may obviously be extended from n-gons to n-arcs (although the notion of pattern does not appear to be of interest for n-arcs). Not every n-arc has the same selfintersection-pattern as a suitable n-gon. For example, the selfintersection-pattern of the 4-arc in Figure 12 cannot be realized by any quadrangle; it is the only selfintersection-pattern of 4-arcs with that property. It is not known how many distinct selfintersection-patterns of ordinary 5-arcs are there.

It is obvious that two n-arcs that have the same selfintersection-pattern may be continuously deformed into each other through n-arcs of the same pattern. Therefore Conjecture 3 is equivalent to the following: If two ordinary n-gons are deformable into each other via n-arcs of the same selfintersection-pattern, they are so deformable via n-gons.

(5)   The chemist Sachse [1890], [1892] appears to have been
the first to consider skew equilateral and isogonal polygons, and
in particular hexagons.   The noteworthy contributions of theoretical chemists continue to this day; see, for example, Dunitz-Waser
[1972*], where the existence of movable 2-equilateral polygons in
E^3 is related to the question of rigidity of polyhedra.

(6)   We call a polygon P in E^d affinely regular provided
for every two flags of P there exists an affine map of P onto
itself that maps the first flag onto the second.   Proposition 10
would lead to a characterization of affinely regular polygons if
the following conjecture were established:

Conjecture 7.   Each affinely regular polygon is a non-singular
affine image of a regular polygon.

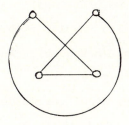

Figure 12

The analogous fact concerning regular convex polytopes was
established by McMullen [1968].

(7)   The notion of regular polygons may be investigated as
well in other geometries (projective, elliptic, spherical, hyperbolic), and in various dimensions.   While in some cases it is not

hard to predict what the outcome will be (for example, the regular skew decagons in Figure 11 lead to regular pentagons in the elliptic plane), in some other cases it is much harder to make a reasonable conjecture.

(8) Another extension of the notion of regularity that remains to be investigated is to regular skew polyhedra (or higher-dimensional objects), having as faces regular skew polygons. We may say that a regular skew polyhedron P is the figure formed (in a suitable space) by a family $\{P_i | i \in I\}$ of regular skew polygons satisfying the usual conditions (two polygons either share precisely one vertex, or one edge, or else have no common elements; the polygons sharing a vertex form a circuit, etc.) such that the symmetries of P act transitively on the flags of P (where a flag of P consists of a polygon P_i, an edge of P_i, and a vertex of that edge). There are no finite regular skew polyhedra that have non-planar regular polygons as faces, but there do exist infinite ones. For example (see the similar construction in Burt [1966, p.39]) starting from the regular square (or hexagonal) tessellation of the plane and raising or lowering alternate vertices, we obtain an infinite regular skew polyhedron consisting of regular nonplanar quadrangles (hexagons). Another example may be obtained from the "regular skew polyhedron 4,6 4 of Petrie" (Coxeter [1937, p.34]) by taking every other vertex as the "center" of a regular skew hexagon; in this polyhedron six equilateral hexagons, isogonal with angles of 60°, meet at each vertex. Clearly, many aspects of this field deserve a more complete treatment. Closely related problems are treated in Schoen [1968] and in Schoen's later abstracts.

(9) We call a polygon P in E^d transitive provided for each pair x_i, x_j of vertices of P there exists an isometry of P that maps x_i onto x_j. A complete classification of transitive polygons seems not to be available even for E^3.

(10)   It is not immediately clear how the second part of Conjecture 6 should be modified for even $n \geq 2k + 2 \geq 6$.

(11)   Generalizing and making more precise the questions treated in Conjectures 5 and 6 and in Theorems 3 and 4, we may ask for the characterization of the sets $\Gamma(n,k,d) \subset E^k$ that consist of all k-tuples $(c_1, c_2, \ldots, c_k)$ possible as parameters for k-equilateral n-gons that span E^d. For example, $\Gamma(5,2,2) = \{ (c_1, (\sqrt{5} - 1)c_1/2),$ $(c_1, (\sqrt{5} + 1)c_1/2) | c_1 > 0 \}$, while $\Gamma(5,2,3) = \emptyset$ and $\Gamma(5,2,4) = \{ (c_1, c_2) | 0 < (\sqrt{5} - 1)c_1/2 < c_2 < (\sqrt{5} + 1)c_1/2 \}$.

REFERENCES

A.D. Aleksandrov

1950   Convex polyhedra. [In Russian]
        Moscow 1950.
        (German translation: A.D. Alexandrow, Konvexe Polyeder,
        Akademie-Verlag, Berlin 1958).

V.I. Arnold

1957   Problem 7. [In Russian]
        Matem. Prosvešč 2 (1957), 268.

A. Auric

1911   Question 3867.
        Intermed. Math. 18 (1911), 122.

L.M. Blumenthal

1953   Theory and applications of distance geometry, Clarendon
        Press, Oxford, 1953.

O. Bottema

1973   Pentagons with equal sides and equal angles, Geometriae
        Didicata 2 (1973), 189-191.

M. Brückner

1900   Vielecke und Vielflache, Teubner, Leipzig 1900.

M. Burt

1966   Spatial arrangement and polyhedra with curved surfaces and
        their architectural application, M. Sc. thesis, Technion-
        Israel Institute of Technology, Haifa, 1966.

C. Carathéodory

1911   Über den Variabilitätsbereich der Fourierschen Konstanten
        von positiven harmonischen Funktionen, Rend. Circ. Mat.
        Palermo 32 (1911), 193-217.

R. Courant and H. Robbins

1941   What is Mathematics? Oxford Univ. Press, New York, 1941.

H.S.M. Coxeter

1937   Regular skew polyhedra in three and four dimensions and
        their topological analogues, Proc. London Math. Soc. (2)
        43 (1937), 33-62.

1974   Regular Complex Polytopes, Cambridge Univ. Press, 1974.

J.D. Dunitz and J. Waser

1972   The planarity of the equilateral, isogonal pentagon, Elem.
        Math. 27 (1972), 25-32.

1972* Geometric constraints in six-and eight-membered rings, J. Amer. Chem. Soc. 94 (1972), 5645-5650.

P. DuVal

1971 Polygon
Encyclopedia Britannica, Chicago, 1971.
Vol. 18, pp.199-201.

V.A. Efremovič and Yu.S. Ilyašenko

1962 Regular polygons in E^n. [In Russian]
Vestnik Moskov. Univ. 1962, No. 5, pp.18-24.

G. Feigl

1925 Über die elementaren Anordnungssätze der Geometrie, J.-Ber. Deutsch. Math.-Verein. 33 (1925), 2-24.

D. Gale

1963 Neighborly and cyclic polytopes, Proc. Symp. Pure Math., Vol. 7 (Convexity); Amer. Math. Soc., Providence, RI, 1963, pp.225-232.

A.P. Garber, V.I. Garvackii and V. Ya. Yarmolenko

1961 Matem. Prosvešč. 6 (1961), pp.345-347.

A. Girard

1626 Tables des sines, tangentes & secantes, selon le raid de 100000 parties. Avec un traicté succinct de la trigonometrie tant des triangles plans, que sphericques. Où sont plusieurs operations nouvelles, non auparavant mises en lumiere, tres-utiles & necessaires, non seulement aux apprentifs; mais aussi aux plus doctes practiciens des mathematiques. Elzevier, La Haye 1626.

S. Günther

1876 Vermischte Untersuchungen zur Geschichte der mathematischen Wissenschaften. Teubner, Leipzig, 1876.

H. Hahn

1908 Über die Anordnungssätze der Geometrie, Monatschefte Math. Phys. 19 (1908), 289-303.

H. Irminger

1970 Zu einem Satz über räumliche Fünfecke, Elem. Math. 25 (1970), 135-136.

S.B. Jackson

1968 A development of the Jordan curve theorem and the Schoenflies theorem for polygons, Amer. Math. Monthly 75 (1968), 989-998.

F. Kárteszi

    1973   Contributo al pentagono equilatero ed isogonale, Ann. Univ.
            Sci. Budapest. Roland. Eötvös Sect. Math. 16 (1973), 63-64.

M.S. Klamkin and B.R. Taylor

    1970   Problem E. 2214
            Amer. Math. Monthly 77 (1970), 79.
            Solution, ibid. pp.1111-1112.

K. Knopp

    1945   Theory of functions, Part 1, Translated from the 5th
            German edition by F. Bagemihl, Dover, New York, 1945.

K. Lamotke

    1958   Der Jordansche Polygonsatz in der affinen Geometrie,
            Enseign. Math. (2) 4 (1958), 272-281.

J. Lawrence

    1975   k-equilateral (2k + 1)-gons span only even-dimensional
            spaces. This volum, pp.

N.J. Lennes

    1911   Theorems on the simple finite polygon and polyhedron, Amer.
            J. Math. 33 (1911), 37-62.

F.J.E. Lionnet

    1873   Note sur une question de géométrie élémentaire, Nouv.
            Annal. de Math. (2) 13 (1873), 331-334.

W. Lüssy and E. Trost

    1970   Zu einem Satz über räumliche Fünfecke, Elem. Math. 25
            (1970), 82-83.

P. McMullen

    1968   Affine and projectively regular polytopes, J. London Math.
            Soc. 43 (1968), 755-757.

A.L.F. Meister

    1769   Generalia de genesi figurarum planarum et inde pendentibus
            earum affectionibus, Novi Comm. Soc. Reg. Scient. Gotting.
            1 (1769/70), pp.144-

A.F. Möbius

    1865   Ueber die Bestimmung des Ihhaltes eines Polyéders, Ber.
            Verh. Königl. Sächs. Ges. Wiss. math.-phys. Kl. 17 (1865),
            31-68 (= Ges. Werke, Vol. 2, pp.473-512, Hirzel, Leipzig
            1886.)

G.A. Pfeiffer

    1937   Polygons.
           Encyclopedia Britannica 14th Edition, New York, Longon,
           Chicago, 1937, Vol. 18, pp.183-186.

L. Poinsot

    1810   Mémoire sur les polygones et les polyèdres, J. École
           Polytech. 10 (1810), 16-48.

V. Polák

    1960   About certain transformations of the simple plane polygons.
           [In Czech; Russian and English summaries] Mat.-Fyz. Časopis
           Slovenk. Akad. Vied. 10 (1960), 81-98.

    1962   On a certain transformation of simple polygonal lines in
           the plane. [In Russian; English summary] Mat.-Fyz. Časopis
           Slovensk. Akad. Vied. 12 (1962), 145-166.

    1968   Mathematical Politology
           Research Memorandum No. 8, Faculty of Science, University
           J.E. Purkyně, Brno 1968, 80 pp.

    1969   One theorem about two parallel simple polygonal lines,
           Spisy Přírod. Fak. Univ. Brno 1969, Číslo 500, 45-50.

A.H. Schoen

    1968   Infinite regular warped polyhedra (IRWP) and infinite
           minimal surfaces (IPMS). Abstract 658-30. Notices Amer.
           Math. Soc. 15 (1968), 727.

H. Sachse

    1890   Über die geometrischen Isomerien der Hexamethylenderivate,
           Ber. Deutsch. Chem. Gesellschaft 23 (1890), 1363-1370.

    1892   Über die Konfigurationen der Polymethylenringe, Zeitschrift
           für physikalische Chemie 10 (1892), 203-241.

M. Simon

    1906   Über die Entwicklung der Elementar-Geometrie im XIX
           Jahrhundert, Jber. Deutsch. Math.-Verien., Ergänzungsband
           1 (1906), viii + 278 pp.

N.J.A. Sloane

    1973   A handbook of integer sequences, Academic Press, New York,
           1973.

S. Šmakal

    1972   Eine Bemerkung zu einem Satz über räumliche Fünfecke, Elem.
           Math. 27 (1972), 62-63.

E. Steinitz

    1916   Polyeder und Raumeinteilungen, Enzykl. math. Wiss. 3 (1922),
           Geometrie, Part 3AB12, pp.1-139.

T.O. Strommer

   1975   Arrangements of lines generated by polygons, Ph.D. thesis,
          University of Washington, Seattle, 1975.

G. Valiron

   1911   Polygones réguliers gauches, Intermed. Math. 18 (1911),
          262.

B.L. van der Waerden

   1970   Ein Satz über räumliche Fünfecke, Elem. Math. 25 (1970),
          73-78.

   1972   Nachtrag zu "Ein Satz über räumliche Fünfecke", Elem. Math.
          27 (1972), 63.

J. Waser

   1944   Ph.D. thesis, Calif. Institute of Techn., Pasadena, 1944.

J. Waser anf V. Schomaker

   1945   On arsenomethane, J. Amer. Chem. Soc. 67 (1945), 2014-2018.

k-EQUILATERAL (2k + 1)-GONS SPAN
ONLY EVEN-DIMENSIONAL SPACES

Jim Lawrence
University of Washington

In his paper [4] in this book, Grünbaum conjectures that no k-equilateral (2k + 1)-gon spans an odd-dimensional space. Here we prove this. We use the notation developed in Grünbaums's paper.

Theorem. Let $P = [x_1, x_2, \ldots, x_{2k+1}]$ be a k-equilateral (2k + 1)-gon affinely spanning E^d. Then d is even.

Proof. Assume that d is odd. We may also assume, since translation preserves the k-equilateral property of the polygon, that $\sum_{i=1}^{2k+1} x_i = 0$.

Since P is k-equilateral, the mapping on the set $S = \{x_1, x_2, \ldots, x_{2k+1}\}$ that takes x_i to x_{i+1} for each i is distance-preserving; i.e., it is a congruence (see Blumenthal [2]). It can be extended to a motion $A : E^d \to E^d$. This motion is an affine mapping, so it preserves $\sum_{i=1}^{2k+1} x_i/(2k+1)$, the origin. Therefore it is a linear mapping. Furthermore, A^{2k+1} fixes each point of S, so it is the identity mapping on E^d.

Since A is a linear motion it preserves the unit sphere. Since this sphere has even dimension there must be a point u of the sphere such that $A(u) = u$ or $A(u) = -u$ (see Aleksandrov [1]). Since $A^{2k+1}(u) = u$, it must be the case that $A(u) = u$.

Let Q be the polytope (see Grünbaum [3]) which is the convex hull of the set S. Its centroid, the origin, is in its interior. Let v be the point on the boundary of Q that is on the ray from the origin through u. Then $A(v) = v$.

Now v is in the relative interior of a unique face of Q. The subset T of S of vertices of this face must be preserved by A, but the only subsets of S preserved by A are S itself and the empty set, neither of which could be the set of vertices of a proper face of Q. Therefore it must be the case that d is even.

REFERENCES

[1]   P.S. Aleksandrov.   Combinatorial Topology, Vol. 3, Craylock Press,
      Albany, N.Y., 1960.

[2]   L.M. Blumenthal.   Theory and Applications of Distance Geometry,
      Clarendon Press, Oxford, 1953.

[3]   B. Grünbaum.   Convex Polytopes, Interscience, London, 1967.

COVERING SPACE WITH CONVEX BODIES

Gulbank D. Chakerian
University of California, Davis

ABSTRACT

Given a family of convex bodies in Euclidean n-space $\mathbb{R}^n$, we consider the problem of finding necessary and sufficient conditions that they can be arranged to cover $\mathbb{R}^n$. Necessary conditions are obtained from variants of the Plank Theorem of Bang. Sufficient conditions are not difficult to obtain, but the main problem is unsolved for $n \geq 3$. Necessary and sufficient conditions in case $n = 2$, established in collaboration with H. Groemer, will be treated in a separate publication. An interesting related unsolved problem is to find the best possible results generalizing the Plank Theorem. Another related problem is that of determining maximal boxes contained in a convex body of given volume. A new proof is given that any n-dimensional convex body of volume V contains a box of volume V/n^n.

COVERING SPACE WITH CONVEX BODIES

Gulbank D. Chakerian
University of California, Davis

Let $\{K_i\}$, $i = 1, 2, \ldots,$ be a family of convex bodies in Euclidean n-space $\mathbb{R}^n$. We shall say that this family can cover $\mathbb{R}^n$ if there exist rigid motions $\{\tau_i\}$, $i = 1, 2, \ldots,$ such that $\bigcup_{i=1}^{\infty} \tau_i K_i = \mathbb{R}^n$. In other words, we require that each K_i can be moved to a position such that the resulting family covers $\mathbb{R}^n$. The main problem we consider, unsolved except in case $n = 1, 2,$ is <u>to find necessary and sufficient conditions that a given family can cover</u> $\mathbb{R}^n$.

We first consider some necessary conditions that must be satisfied if $\{K_i\}$ can cover $\mathbb{R}^n$. Let $V(K)$ denote the n-dimensional content of K. If $\{K_i\}$ can cover $\mathbb{R}^n$, then certainly we must have

$$(1) \quad \sum V(K_i) = \infty.$$

Let $w(K, u)$ be the distance between the supporting hyperplanes of K orthogonal to the direction u, and let $w(K) = \min w(K, u) =$ the minimum width of K. The Plank Theorem of Bang [1] implies that if an n-dimensional convex body K is covered by convex bodies $\{K_i\}$, then $\sum w(K_i) \geq w(K)$. Hence if the family $\{K_i\}$ can cover $\mathbb{R}^n$, then we must have

$$(2) \quad \sum w(K_i) = \infty.$$

The conditions (1) and (2), although necessary, are not sufficient that $\{K_i\}$ can cover $\mathbb{R}^n$, even in case $n = 2$. For example, consider in $\mathbb{R}^2$ the family consisting of $\frac{1}{m} \times \frac{1}{m}$ squares, $m = 1, 2, \ldots,$ together with $\frac{1}{m^2} \times m^2$ rectangles, $m = 1, 2, \cdots$ It is easy to show that the rectangles must leave uncovered area at least $\pi r^2 - \pi^2 r / 3$ of the area circular disk of radius r (simply examine the maximum area of the intersection of an infinite strip of given width with a circular disk of given radius), and this remainder cannot be covered by the squares if r is sufficiently large. Hence this family cannot cover $\mathbb{R}^2$, although it satisfies both conditions (1) and (2).

A sufficient condition for covering, in the case of $\mathbb{R}^2$, is provided by the following theorem.

Theorem 1. Let $\{K_i\}$ be a family of plane convex bodies whose areas are bounded below by a positive constant, and such that $\sum w(K_i) = \infty$. Then the family can cover $\mathbb{R}^2$.

Proof. Any plane convex body K of area A and minimum width w contains a rectangle of area at least A/4 and having each side at least w/2 in length (see [4], or the Box Lemma at the end of this article). This rectangle has longest side at least $(\sqrt{A})/2$ in length. Thus the conditions on the given family $\{K_i\}$ imply there is a constant $\lambda > 0$ such that each K_i contains a rectangle R_i with one side of length λ and the other of length $w(K_i)/2$. Now tile the plane with $\lambda \times \lambda$ squares in the usual way and let these squares be enumerated $Q_1, Q_2, \cdots$. From the condition $\sum w(K_i) = \infty$ it is obvious how the squares can be successively covered by the rectangles, and hence $\mathbb{R}^2$ can be covered by $\{K_i\}$. This completes the proof.

Another useful sufficient condition in $\mathbb{R}^2$ is provided by the following theorem, whose proof we omit.

Theorem 2. Let $\{K_i\}$ be a family of plane convex bodies whose diameters are bounded above, and such that $\sum V(K_i) = \infty$. Then the family can cover $\mathbb{R}^2$.

Both the preceding theorems are in fact consequences of the following theorem, established in collaboration with H. Groemer and whose proof will appear in a joint publication.

Theorem 3. Let $\{K_i\}$ be a family of plane convex bodies with areas $V(K_i)$ and diameters $D(K_i)$. Then $\{K_i\}$ can cover $\mathbb{R}^2$ if and only if

$$\sum \frac{V(K_i)}{1 + D(K_i)} = \infty.$$

The situation in higher dimensions is considerably more complex. For a convex body K in $\mathbb{R}^n$ let $\sigma(K,u)$ denote the $(n-1)$-dimensional content of the orthogonal projection of K onto a hyperplane orthogonal to the direction u, and let $\sigma(K) = \min \sigma(K,u)$. Then a necessary condition that a family $\{K_i\}$ can cover $\mathbb{R}^n$ is

$$(3) \quad \sum \sigma(K_i) = \infty.$$

This follows from a variant of the Plank Theorem asserting that if $\{K_i\}$ covers K in $\mathbb{R}^n$ then

$$(4) \quad \sum \sigma(K_i) \geq \frac{1}{n} \sigma(K).$$

The conditions (1), (2), and (3), although necessary, are not sufficient that $\{K_i\}$ can cover $\mathbb{R}^n$, even in case $n = 3$. For example, consider in $\mathbb{R}^3$ the family consisting of $m \times m \times \frac{1}{m^2}$ boxes together with $\frac{1}{m} \times \frac{1}{m} \times m^2$ boxes, $m = 1, 2, \cdots$. An analysis similar to that done previously for the analogous example in $\mathbb{R}^2$ shows that this family cannot cover $\mathbb{R}^3$ (since it can cover only finite volume). Yet conditions (1), (2), and (3) are satisfied. Indeed, note that not only is condition (1) satisfied by this family, but in fact all the sets have unit volume!

If $\sigma_\ell(K)$ denotes the minimum of the ℓ-dimensional content of the orthogonal projection of K into any ℓ-flat, then $\sigma_1(K) = w(K)$ and $\sigma_{n-1}(K) = \sigma(K)$. It is natural to conjecture necessary and sufficient conditions that $\{K_i\}$ can cover $\mathbb{R}^n$ will involve the $\sigma_\ell(K_i)$ for $\ell = 1, 2, \ldots, n-1$. It would in any case be of independent interest to establish the best possible inequalities analogous to (4), with "σ_ℓ" replacing "σ" and some better constant (if possible) replacing the "$\frac{1}{n}$". It was already shown by Bang [1] that "$\frac{1}{n}$" may not be replaced by "1" in the equality (4).

One would expect that results about boxes inside convex bodies, generalizing the result used in the proof of Theorem 1, might be useful in analyzing the covering problem in $\mathbb{R}^n$. On the other hand, such results are useful in other applications and are of independent interest. It is known that any n-dimensional convex body K contains a rectangular box of volume at least $V(K)/n^n$. A nice inductive proof is given by Hadwiger [2], establishing simultaneously the existence of a circumscribed box of volume at most $(n!)V(K)$. The following lemma gives another proof of the result about enclosed boxes, and shows that all edges of the box may be chosen relatively long. The proof is also of interest because of its application of a topologycal theorem of Rattray.

   Box Lemma. Inside any n-dimensional convex body K there is a rectangular box of volume at least $V(K)/n^n$ and having all edges at least $w(K)/n$ in length.

   Proof. Standard approximation arguments show that it suffices to assume K is smooth and strictly convex. If S^{n-1} denotes the standard unit sphere in $\mathbb{R}^n$, there is associated with each direction $u \in S^{n-1}$ the unique point $x(u)$ on the boundary of K with outward pointing unit normal u. The line segment with endpoints $x(u)$ and $x(-u)$ is a diameter of K. The mapping $T:S^{n-1} \to S^{n-1}$, defined by $T(u) = (x(u) - x(-u))/\|x(u) - x(-u)\|$, $u \in S^{n-1}$, is continuous and sends antipodes to antipodes, that is $T(-u) = -T(u)$. This is precisely the situation needed to apply the theorem of Rattray [3], which asserts then that some orthogonal n-tuple is mapped by T into an orthogonal n-tuple. That is, there exist mutually orthogonal $u_1, u_2, \ldots, u_n \in S^{n-1}$ such that $T(u_1), T(u_2), \ldots, T(u_n)$ are again mutually orthogonal. This shows that K admits a circumscribed box B_1 such that the diameters of K determined by opposite faces of B_1 are mutually orthogonal. The 2^n points of the form $(x(\pm u_1) + x(\pm u_2) + \ldots + x(\pm u_n))/n$ are the vertices of a rectangular box B_0 contained in K and having its edges parallel to the directions $T(u_1), \ldots, T(u_n)$. Each edge of B_0 has length $1/n$ times the length of the diameter of K parallel to that edge. Since all diameters of K have length at least $w(K)$, it follows that each edge of B_0 has length at least $w(K)/n$. Since each edge of B_0 is at least $1/n$ the length of a corresponding edge of B_1, it follows that $V(B_0) \geq V(B_1)/n^n \geq V(K)/n^n$. This completes the proof.

REFERENCES

[1] Th. Bang, On covering by parallel strips, Mat. Tidsskrift B
     (1950), 49-53.

[2] H. Hadwiger, Volumschätzung für die einen Eikörper überdeckenden
     und unterdeckenden Parallelotope, Elem. d. Math. 10 (1955),
     122-124.

[3] B.A. Rattray, An antipodal-point, orthogonal point theorem, Ann.
     of Math. 60 (1954), 502-512.

[4] Amer. Math. Monthly 80 (1973), 562-563, Solution to Problem
     E2360 [1972, 519].

INTERSECTIONAL CONFIGURATION OF CONVEX SETS

Clinton M. Petty
University of Missouri

Dedicated to Leonard M. Blumenthal

ABSTRACT

For brevity, we call a family $\mathcal{J} = \{K_\alpha | \alpha \in I\}$ of closed convex sets, in d-dimensional euclidean space E^d, an intersectional configuration of class n if $\mathcal{J}$ satisfies the following conditions:

(i) The intersection $K = \cap \{K_\alpha | \alpha \in I\}$ is a convex body set with a nonempty interior) and K is a proper subset of each K_α.

(ii) There exists an integer $m \geq 2$ such that the intersection of any m members of $\mathcal{J}$ is K. The class n is the smallest such value of m.

L.M. Blumenthal (private communication) pointed out to the author that, for $d = n = 2$, an intersectional configuration $\mathcal{J}$ with three members has the property that if a convex set B intersects all three members of $\mathcal{J}$ then B must intersect K. In this paper, we extend this result to the general case by proving the following theorem.

Theorem. If a convex set B intersects $dn - d + 1$ or more members of an intersectional configuration $\mathcal{J}$, then B intersects K.

Examples are given which show that, for every d and every n, this is the best result possible. The method of proof is based in some results of Minkowski and Favard and involves a novel but interesting classification of supporting hyperplanes to a convex body.

Numerous examples of intersectional configurations have led

the author to the following conjecture:

Conjecture. If $\mathcal{J} = \{K_\alpha | \alpha \in I\}$ is an intersectional configuration of class n in E^d, then $\cup \{K_\alpha | \alpha \in I\}$ can be expressed as a union of $(d + 1)(n - 1)$ or fewer closed convex sets.

In support of this conjecture, we show that it is true for all n if $d = 2$ and in addition for each n this is the best result possible.

INTERSECTIONAL CONFIGURATIONS OF CONVEX SETS

Clinton M. Petty
University of Missouri

Dedicated to Leonard M. Blumenthal

1. Introduction

The following observation of L.M. Blumenthal (private communication) motivated this paper. If three convex figures in the plane have the property that the intersection of each pair is the same convex figure K, then any convex set which intersects all three must also intersect K. Three dominoes, so situated that the intersection of each pair is a half-domino, is an example of such an intersectional configuration.

This observation suggests that such configurations possess interesting geometrical properties. Accordingly, our first task is to generalize this concept. For brevity, we call a family $\mathcal{J} = \{K_\alpha | \alpha \in I\}$ of closed convex sets in d-dimensional euclidean space E^d, an intersectional configuration of class n if $\mathcal{J}$ satisfies the following conditions:

(i) The intersection $K = \cap \{K_\alpha | \alpha \in I\}$ is a convex body (compact convex set with a nonempty interior) and K is a proper subset of each K_α.

(ii) There exists an integer $m \geq 2$ such that the intersection of any m members of $\mathcal{J}$ is K. The class n is the smallest such value of m.

It may be observed that the index set I is countable (finite or denumerable). For, if S is a denumerable dense subset of the (open) complement $E^d \sim K$ of K, then each point of S belongs to at most $n - 1$ members of $\mathcal{J}$ and each K_α intersects S since K_α has interior points in $E^d \sim K$. A subfamily of $\mathcal{J}$ with n or more members is also an intersectional configuration with the same intersection K but, perhaps, of a lower class.

Our first goal will be to establish (Section 2) the following generalization of the observation of Blumenthal which corresponds to the case $d = n = 2$.

Theorem 1. If a convex set B intersects $dn - d + 1$ or more members of an intersectional configuration $\mathcal{J}$, then B intersects K.

We will also show that, for every d and every n, this is the best result possible (Section 3).

Certain intersectional configurations have already been studied in detail in the theory of convex sets. If p is an exterior point of a convex body K, then the convex hull $K(p)$ of K and p is called a simple Kappenkörper of K. Every supporting plane of $K(p)$, with the exception of those which touch $K(p)$ only in the corner p, is also a supporting plane of K. A general Kappen-körper of K is a convex body $K_1 \supset K$ such that every supporting plane of K_1, with the exception of those which touch K_1 only in a corner, is also a supporting plane of K. A well-known result (see [1, pp.17-18]) states that the general Kappenkörper of a convex body K is the union of countably many simple Kappenkörper and the intersection of any two of them is K itself. Thus the family of these simple Kappenkörper is an intersectional configuration of class 2.

In one respect, this example is not typical of an intersectional configuration. For the union $F = \cup \{K_\alpha | \alpha \in I\}$ is not in general a convex set. However, there appears to be a natural restriction on just how nonconvex F can be. In this direction, we offer the following conjecture:

Conjecture. The union $F = \cup \{K_\alpha | \alpha \in I\}$ of the members of an intersectional configuration can be expressed as the union of $(d + 1)(n - 1)$ or fewer closed convex sets.

In support of this conjecture, we will show (Section 4) that it is true for all n if d = 2 and in addition for each n this is the best result possible.

2. Intersection Properties

A study of examples of intersectional configurations reveals that a supporting plane of the intersection K can be a separating plane of at most a certain number of the $\{K_\alpha\}$ and this number depends on a special classification of supporting planes.

With the origin at an interior point of K, let H(u) be the supporting function of K and let K^* be the polar reciprocal of K with respect to the unit sphere Ω. We are led to the following ad hoc classification of boundary points and supporting planes.

Definition. A boundary point p of K is said to be of order k if k is the dimension of a minimal simplex which contains p and whose vertices are extreme points of K. A supporting plane to K with outer normal $u \in \Omega$ is said to be of order k if the boundary point $H^{-1}(u)u$ of K^* is of order k.

As usual, the extreme points and the extreme supporting planes are of order 0. However, in some respects this classification is unusual. For example, the boundary points of order 1 of a cube are those points which are inner points of an edge or a face diagonal. Thus, in the dual figure (an octahedron) the supporting planes which touch the octahedron only in a vertex do not all have the same order.

Since K = conv(ext K), it follows from Caratheodory's Theorem that $0 \leq k \leq d - 1$. Let

$$L(u,x) = \langle u,x \rangle - H(u), u \in \Omega, x \in E^d.$$

If the supporting plane to K with equation L(u,x) = 0 is of order k, then

$$(1) \quad L(u,x) = \sum_{i=0}^{k} u_i L(u_i,x), \quad x \in E^d, \quad \mu_i > 0 (i = 0, \ldots, k),$$

where each $L(u_i,x) = 0$ is the equation of a supporting plane to K of order 0.

Theorem 2. Let $\mathcal{J} = \{K_\alpha | \alpha \in I\}$ be an intersectional configuration of class n in E^d. If H is a supporting plane to $K = \cap \{K_\alpha | \alpha \in I\}$ of order k, then H is a separating plane of at most $(k+1)(n-1)$ members of $\mathcal{J}$.

Proof. Let $H = \{x | L(u,x) = 0\}$ be an extreme supporting plane to K. Suppose $L^+(u) = \{x | L(u,x) > 0\}$ intersects (at least) n members of $\mathcal{J}$, say $K_1, \ldots, K_n$. Then, there exists a hyperplane $H_1 \subset L^+(u)$ parallel to H at distance $c > 0$ so that H_1 intersects each K_i ($i = 1, \ldots, n$). Let $p_i \in H_1 \cap K_i$ and put $M = \max \|p_i - p_j\|$. On the negative side of H, the hyperplane parallel to H at distance $\delta > 0$ will intersect K (for sufficiently small δ) in a section which contains a maximal $(d-1)$-dimensional ball B_{d-1} with center x and radius $\rho(\delta)$. Now, the condition $\delta^{-1}\rho(\delta) \to \infty$ as $\delta \to 0$ characterizes extreme supporting planes. This characterization is due to H. Minkowski [4, Ges. Abh. pp.166-168] for $d = 3$ and extended to the general case by J. Favard [3, pp.230-232]. Now, choose δ sufficiently small so that $\delta^{-1}\rho(\delta) > c^{-1}M$. Let the cone $C_i \subset K_i$ with apex p_i and base B_{d-1} intersect H in the $(d-1)$-dimensional ball with center q_i and radius r. Then, $q_i = (c + \delta)^{-1}(cx + \delta p_i)$ and $r = (c + \delta)^{-1}c\rho(\delta)$. Consequently, $\|q_i - q_j\| < r$ and each q_i is an interior point of each K_j. In particular, $q_1 \in \text{int}(K_1 \cap \ldots \cap K_n) = \text{int } K$ contrary to the fact that q_1 lies in the supporting plane H of K. The extension to supporting planes of order k now follows from (1).

Corollary 1. Let $\mathcal{J} = \{K_\alpha | \alpha \in I\}$ be an intersectional configuration of class n in E^d and let $K = \cap \{K_\alpha | \alpha \in I\}$.

(a) The union of any subfamily of $\mathcal{J}$ is closed.

(b) There exists an extreme supporting plane to K which is a separating plane for exactly $n - 1$ members of $\mathcal{J}$.

(c) If K is a polytope with f_{d-1} facets, then $\mathcal{J}$ has not more than $(n - 1)f_{d-1}$ members.

Proof. To prove (a), we need only consider a boundary point x of $F = \cup \{K_\alpha | \alpha \in I\}$ which is exterior to K. Then for some $u \in \Omega$, $x \in L^+(u)$ and Theorem 2 implies that $x \in F$.

A proof of (b) follows from the definition of the class n, Theorem 2, and the fact that K is the intersection of all closed halfspaces which contain K and are bounded by extreme supporting planes of K.

To prove (c), one observes that the extreme supporting planes of a convex d-polytope are those hyperplanes which contain a facet.

We can now give a simple proof of Theorem 1. Suppose B intersects, say, $K_1, \ldots, K_s$ where $s = dn - d + 1$. Let $p_i \in B \cap K_i$ and put $C = \operatorname{conv}\{p_1, \ldots, p_s\} \subseteq B$. If B does not intersect K, then C and K may be strictly separated by a hyperplane. It follows that there exists a supporting plane of K which must be a separating plane of each of the convex sets $K_1, \ldots, K_s$ contrary to Theorem 2.

We will show by example in the next section that Theorem 2 gives the best possible result for every k, n and d.

3. A Key Example

Let $\{u_0, \ldots, u_d\}$ be the vertices of a regular d-simplex Δ inscribed in the unit sphere Ω. Let $I = \{(i, j) | i = 0, \ldots, d; j = 1, \ldots, n-1\}$ and let $\{e_{ij} | (i, j) \in I\}$ be $(d + 1)(n - 1)$ distinct points such that e_{ij} lies in the relative interior of the facet of Δ opposite u_i. Then

$$e_{ij} = \sum_{k=0}^{d} \alpha_k^{ij} u_k, \quad \alpha_k^{ij} > 0 \text{ for } k \neq i, \quad \alpha_i^{ij} = 0, \quad \sum_{k=0}^{d} \alpha_k^{ij} = 1.$$

Let $x_{ij} = -(d + 1)u_i + e_{ij}$, $K_{ij} = \text{conv}\{x_{ij}, \Delta\}$ and $\mathcal{J} = \{K_{ij} | (i,j) \in I\}$. Using Barycentric coordinates $(\beta_0, \ldots, \beta_d)$ with respect to the affine basis $\{u_0, \ldots, u_d\}$, we have

$$(2) \quad x_{ij} = \sum_{k=0}^{d} \beta_k^{ij} u_k, \quad \beta_k^{ij} = 1 + \alpha_k^{ij} \quad \text{for} \quad k \neq i, \quad \beta_i^{ij} = -d.$$

Only the i-th Barycentric coordinate of a point in K_{ij} can be negative. Consequently, $K_{ij} \cap K_{km} = \Delta$ for $i \neq k$ and $\mathcal{J}$ is an intersectional configuration of class n with $\Delta = \cap \{K_{ij} | (i,j) \in I\}$.

Let $\{v_i\}$ with $v_i = -du_i$, be the vertices of the polar reciprocal Δ^* of Δ. For a fixed k satisfying $0 \leq k \leq d - 1$, $v = (k + 1)^{-1} \sum_{i=0}^{k} v_i$ is a boundary point of Δ^* of order k and, therefore, the supporting hyperplane to Δ with outer normal $u = v/\|v\|$ is of order k. Calculating $L(u,x)$, we obtain

$$L(u,x) = [(k + 1)\|v\|]^{-1} \sum_{i=0}^{k} (d\langle -u_i, x \rangle - 1),$$
$$\|v\| = [d(d - k)(k + 1)^{-1}]^{1/2}.$$

Since $\langle -u_i, u_j \rangle = d^{-1}$ for $i \neq j$, from (2) we obtain

$$L(u, x_{ij}) = -[(k + 1)\|v\|]^{-1}(d + 1) \sum_{m=0}^{k} \beta_m^{ij}.$$

Consequently, $L^+(u)$ intersects K_{ij} for $i = 0, \ldots, k$ and $j = 1, \ldots, n-1$.

The quantities in Theorem 2 are, therefore, the best possible. With $k = d - 1$, the open halfspace $L^+(u)$ is a convex set which intersects $d(n - 1)$ members of $\mathcal{J}$ but $L^+(u)$ does not intersect Δ. Hence, Theorem 1 cannot be improved.

We observe that the line segment determined by any two points in $\{x_{ij}\}$ does not lie in the union $F = \cup \{K_{ij} | (i,j) \in I\}$. Thus, F cannot be expressed as a union of fewer than $(d + 1)(n - 1)$ convex sets. If the Conjecture stated in the introduction is

correct, then this example is extremal with respect to this property also. In the next section we will show that this is actually
the case when $d = 2$.

4. A Union of Convex Sets

For the convenience of the reader we will divide the proof of
the following result into three Lemmas.

Theorem 3. Let $\mathcal{J} = \{K_\alpha | \alpha \in I\}$ be an intersectional config-
uration of class n in E^2. Then $F = \cup \{K_\alpha | \alpha \in I\}$ can be
expressed as a union of $3n - 3$ or fewer closed convex sets.

We first replace $\mathcal{J}$ by a simpler intersectional configuration
which has the same intersection and union and the same class.

For $x \in K_\alpha \sim K$, let $A_\alpha(x) = \{u \in \Omega | x \in L^+(u)\}$. Each $A_\alpha(x)$
is an open arc of Ω. The union $U_\alpha = \cup \{A_\alpha(x) | x \in K_\alpha \sim K\}$ may be
(uniquely) expressed as a countable family $\{A_\alpha^i | i \in J_\alpha\}$ of pair-
wise disjoint open arcs. For convenience, we consider Ω itself
to be an arc. Let $B_\alpha^i = K \cup \{x \in K_\alpha \sim K | A_\alpha(x) \subset A_\alpha^i\}$.

Lemma 1. The family $\mathcal{B} = \{B_\alpha^i | \alpha \in I, i \in J_\alpha\}$ is an inter-
sectional configuration of class n with intersection K and
union F.

Proof. We first show that each B_α^i is convex. Let $x, y \in B_\alpha^i$
$\subset K_\alpha$ and let z be a point of the segment xy. We may assume
that $z \in K_\alpha \sim K$ and, consequently, for some $u \in \Omega$ we have
$z \in L^+(u)$. Thus, for at least one of the points x or y, say x,
we have $x \in L^+(u)$. Therefore, $u \in A_\alpha(z) \cap A_\alpha(x)$. Since $A_\alpha(x) \subset$
A_α^i, it follows that $A_\alpha(z) \subset A_\alpha^i$ and $z \in B_\alpha^i$. To show that B_α^i
is closed, let y be a boundary point of B_α^i. Since K_α is closed
$y \in K_\alpha$ and we may assume that $y \in K_\alpha \sim K$. Thus, for some $u \in \Omega$,
we have $y \in L^+(u)$. Since y is a boundary point of B_α^i, for some
$x \in B_\alpha^i$ we have $x \in L^+(u)$ and, as above, this implies $y \in B_\alpha^i$.

Next, we show that the intersection of any n members of $\mathcal{B}$

is K. Suppose two of these members have the same subscript, say B_α^i and B_α^j, $i \neq j$. If for some $x \in K_\alpha \sim K$ we have $x \in B_\alpha^i \cap B_\alpha^j$, then $A_\alpha(x) \subset A_\alpha^i \cap A_\alpha^j$ which implies $i = j$. Consequently, $B_\alpha^i \cap B_\alpha^j = K$. On the other hand, if the n members of $\mathcal{B}$ have distinct subscripts, then this property follows from the corresponding property of $\mathcal{J}$.

Finally, we observe that $\mathcal{B}$ is of class n. Since $\mathcal{J}$ is of class n, there exist, say, $K_1, \ldots, K_{n-1}$ and a point $x \in K_1 \cap \ldots \cap K_{n-1}$ such that $x \notin K$. Thus, x belongs to $n-1$ members of $\mathcal{B}$.

Lemma 2. If $\mathcal{A}^* \subset \mathcal{A} = \{A_\alpha^i \mid a \in I, i \in J_\alpha\}$ is an irreducible subcovering of $U = \cup \{A_\alpha^i \mid a \in I, i \in J_\alpha\}$, then the union F^* of the corresponding subfamily $\mathcal{B}^* \subset \mathcal{B}$ can be expressed as the union of three or fewer closed convex sets.

Proof. If three or fewer members of $\mathcal{A}^*$ cover U, then there are no other members of $\mathcal{A}^*$ and the proof is complete. We may assume that each three arcs A_α^i, A_β^j, A_γ^k of $\mathcal{A}^*$ do not cover U. Then a pair of them, say A_α^i and A_β^j, are disjoint. For suppose each pair have a point in common. Then, by Helly's theorem on the line, all three have a point in common and one of them must be contained in the union of the other two. Now, we show that $B_\alpha^i \cup B_\beta^j$ is convex. Let $x \in B_\alpha^i$ and $y \in B_\beta^j$ and let L be the line through x and y. If L intersects K, then the segment xy must belong to $B_\alpha^i \cup B_\beta^j$. If L does not intersect. L, then there exists $u \in A_\alpha^i \cap A_\beta^j$ contrary to the fact that A_α^i and A_β^j are disjoint.

By the above argument, any three points of F^* have the property that the segment joining two of them must lie in F^*. By Corollary 1a, F^* is closed and a result of Valentine [5] implies that F^* may be expressed as the union of three of fewer closed convex sets.

Lemma 3.

(a) If $u \in \Omega$, then not more than $2n - 2$ members of $\mathcal{A}$ contain u.

(b) If u is the outer normal of an extreme supporting line of K, then not more than $n - 1$ members of $\mathcal{A}$ contain u.

(c) For some outer normal u of an extreme supporting line of K, exactly $n - 1$ members of $\mathcal{A}$ contain u.

(d) Each arc A_{α}^{i} contains an outer normal of an extreme supporting line of K.

Proof. A proof of parts (a), (b) and (c) follows from Theorem 2 and Corollary 1b. A proof of part (d) follows from a standard argument.

We shall now prove Theorem 3 by induction on the class n. For $n = 2$, parts (b) and (d) of Lemma 3 imply that $\mathcal{A}$ is an irreducible covering of U and thus, the proof follows from Lemmas 1 and 2.

We may now assume that $n \geq 3$ and that the result is valid for smaller values of the class. By Lemma 3a, $\mathcal{A}$ is a point-finite covering of U and, consequently, there exists an irreducible subcovering $\mathcal{A}^{*}$ of U (see [2, p.160]). Lemma 2 implies that the corresponding $\mathcal{B}^{*}$ can be expressed as the union of three or fewer closed convex sets. Removing $\mathcal{B}^{*}$ from $\mathcal{B}$, there are either fewer than n members remaining or, by Lemma 3c, the class is lowered. In either case, the result follows and the proof is complete.

REFERENCES

[1] T. Bonnesen and W. Fenchel, Theorie der konvexen Körper,
     Ergebnisse der Mathematik, Vol. 3, No. 1, Springer, Berlin,
     1934.

[2] J. Dugundji, Topology, Allyn and Bacon, Boston, 1966.

[3] J. Favard, Sur les corps convexes, J. Math. pures appl. (9)
     12 (1933), 219-282.

[4] H. Minkowski, Theorie der konvexen Körper, insbesondere
     Bergrundung ihres Oberfläche, Math. Ann. 57 (1903), 447-495;
     Ges. Abh. Bd. 2, 230-276.

[5] F.A. Valentine, A three point convexity property, Pacific J.
     Math. 7 (1957), 1227-1235.

METRIC DEPENDENCE AND A SUM OF DISTANCES

Dorothy Wolfe
Widener College

ABSTRACT

In a real normed linear space B a point q is said to be metrically dependent on $\{p_1, \ldots, p_n\}$ if there exists a set of weights $\{a_i\}$ (non-negative numbers with sum 1) such that for each w in B $d(w,q) \leq \sum_1^n a_i d(w,p_i)$.

Theorem 1. q lies inside the convex cover of a set if and only if it is metrically dependent on the set.

Theorem 2. Given a point q, if there exists a set of weights such that the weighted average of distances from q to $p_1, \ldots, p_n$ is greater than the average with the same weights of distances from p_j to $p_1, \ldots, p_n$ for each j, then q is not metrically dependent on the set. That is q lies outside its convex cover.

These theorems are used to prove

Theorem 3. In any real normed linear space, given r points on the boundary of the unit sphere, if their convex cover contains the origin, the sum of their distances is at least $2(r - 1)$.

METRIC DEPENDENCE AND A SUM OF DISTANCE

Dorothy Wolfe
Widener College

Information about convexity properties of sets can be squeezed from metric properties, that is from a knowledge of distances between points. These convexity properties, in turn, affect distances. We shall use these two ideas to calculate the minimum of a sum of certain distances on the unit sphere.

Definition. Let $\{a_i\}$ be a set of weights if $a_i \geq 0$ for every i and $\sum_1^n a_i = 1$. Then if q, $p_1, \ldots, p_n$ are members of a real normed linear space B, we say that q is metrically dependent on $\{p_1, \ldots, p_n\}$ if there exists a set of weights $\{a_i\}$ such that for each $w \in B$

$$d(w,q) \leq \sum_1^n a_i d(w,p_i)$$

Theorem 1. q lies inside the convex cover of $\{p_1, \ldots, p_n\}$ if and only if q is metrically dependent on the set.

(For proofs of this and the following theorem see 2.) Theorem 1 involves each element of the space, but Theorem 2 is concerned with only the finite set of points:

Theorem 2. If there exists a set of weights $\{b_i\}$ such that $\sum_{i=1}^n b_i d(q,p_i) > \sum_{i=1}^n b_i d(p_j,p_i)$ for each j, then q is not metrically dependent on $\{p_1, \ldots, p_n\}$.

In other words, if some weighted average of distance from q to the p_j's is greater than the average with the same weights of distances from each p_j to the p_i's, then q must be an extreme point of the convex cover of the set $\{p_1, \ldots, p_{n-q}\}$ whenever it is embedded in any normed linear space.

Using a bit of arithmetic on the weights and letting q vary in the convex cover of $\{p_1, \ldots, p_n\}$, we can express Theorem 2 in the following form:

Corollary. In any real normed linear space, given a convex polytope P, the convex cover of $\{p_1, \ldots, p_n\}$, if $\{c_i\}$ is any set of non-negative numbers, $\max\limits_{q \in P} \sum\limits_{i=1}^{n} c_i d(q, p_i)$ is attained when q is one of the extreme points of P.

We use this concept to prove a conjecture of Chakerian and Klamkin on a sum of distances on the unit sphere: (See 1, where the theorem is proved for Euclidean spaces and for the Minkowski plane.)

Theorem 3. In any real normed linear space, given r points on the boundary of the unit sphere, if their convex cover contains the origin, the sum of their distances is at least $2(r-1)$.

This minimum is the best possible. One can approach it arbitrarily closely by taking one point at the north pole of the sphere and the others clustered near the south pole. The proof is related to this idea, using the corollary to Theorem 2 to pick out the north pole:

Lemma. With conditions of Theorem 3, there is at least one point -- say p_1 -- such that $\sum\limits_{i=1}^{r} d(p_1, p_i) \geq r$.

Proof of Lemma. Since the origin, O, lies inside the convex cover of $\{p_1, \ldots, p_r\}$ we use the corollary, setting each $c_i = 1$. Then since

$$\sum_{i=1}^{r} d(O, p_i) = r$$

there exists j between 1 and r such that

$$(1) \quad \sum_{i=1}^{r} d(p_j, p_i) \geq r$$

Proof of Theorem. If some other distance, for instance, $d(p_2, p_3) \geq 1$, the lemma is enough to prove the theorem. For then

$$(2) \quad d(p_2, p_i) + d(p_3, p_i) \geq 1 \quad i = 4, \ldots, r \quad \text{and}$$

$$d(p_2, p_3) + \sum_{i=4}^{r} [d(p_2, p_i) + d(p_3, p_i)] \geq r - 2$$

and adding (1) and (2) we're finished.

If $d(p_i, p_j) < 1$ for all $i, j > 1$, let $\max\limits_{i,j>1} d(p_i, p_j) = d(p_2, p_3) <$

Then, by the same use of the triangle inequality as above in (2)

$$\sum_{1<i<j} d(p_i, p_j) \geq (r - 2)d(p_2, p_3)$$

and we need only show

(3) $$\sum_{i=2}^{r} d(p_1, p_i) \geq 2(r - 1) - (r - 2)d(p_2, p_3).$$

Again we use the corollary to Theorem 2. We set

$$c_1 = (r - 1) - (r - 2)d(p_2, p_3) > 0$$

$$c_i = 1 \qquad\qquad\qquad i = 2, \ldots, r$$

Then

$$\sum_{i=1}^{r} c_i d(p_i, 0) = 2(r - 1) - (r - 2)d(p_2, p_3)$$

Therefore there exists j such that

$$\sum_{i=1}^{r} c_i d(p_j, p_i) \geq 2(r - 1) - (r - 2)d(p_2, p_3)$$

But for $j \neq 1$

(4) $$\sum_{i=1}^{r} c_i d(p_j, p_i) \leq 2c_1 + (r - 2)d(p_2, p_3)$$

$$= 2[(r - 1) - (r - 2)d(p_2, p_3)] +$$
$$(r - 2)d(p_2, p_3)$$
$$= 2(r - 1) - (r - 2)d(p_2, p_3)$$

Equality cannot hold in (4) unless (among other conditions) $d(p_j, p_1) = 2$. In this case, triangle inequalities are sufficient to prove inequality (3) and the theorem.

If the inequality (4) is strict, j can only be 1 and, indeed,

$$\sum_{i=1}^{r} c_i d(p_1, p_i) = \sum_{i=2}^{r} d(p_1, p_i)$$

$$\geq 2(r - 1) - (r - 2)d(p_2, p_3).$$

REFERENCES

[1] G.D. Chakerian and M.S. Klamkin, *Inequalities for sums of distances*, Am. Math. Monthly 80 (1973), 1009-1017.

[2] Dorothy Wolfe, *Metric inequalities and convexity*, Proc. Amer. Math. Soc. 40 (1973), 559-562.

TVERBERG-TYPE THEOREMS IN CONVEX PRODUCT STRUCTURES

Gerald Thompson
Augusta College

William R. Hare
Clemson University

ABSTRACT

Corresponding to the work of Eckhoff in studying Radon's theorem in an abstract setting, the present paper examines the abstract analogue of Tverberg's generalization of Radon's theorem. (Tverberg's theorem: Each set of $(j - 1)(d + 1) + 1$ points in R^d can be partitioned into j nonempty subsets whose convex hulls have a point in common.) A convex structure $(X, \mathcal{C}(X))$ consists of a nonempty set X and a family $\mathcal{C}(X)$ of subsets of X satisfying (i) X and $\emptyset$ are members of $\mathcal{C}(X)$ and (ii) if $\mathcal{J}$ is any subfamily of $\mathcal{C}(X)$, then $\cap \mathcal{J}$ is a member of $\mathcal{C}(X)$. For two convex structures $(X, \mathcal{C}(X))$ and $(Y, \mathcal{C}(Y))$, the product structure is $(X \times Y, \mathcal{C}(X) \oplus \mathcal{C}(Y))$, where $\mathcal{C}(X) \oplus \mathcal{C}(Y) = \{A \times B : A \in \mathcal{C}(X), B \in \mathcal{C}(Y)\}$. The convex hull of any subset A of X is defined to be the intersection of the family of elements of $\mathcal{C}(X)$ which contain A. The j-order Radon index, $r_j(X, \mathcal{C}(X))$, is the smallest nonnegative integer r (if such exists) such that, if A is a subset of X having at least $r + 2$ elements, then A can be partitioned as $A = \bigcup_{i=1}^{j} A_j$ with $\bigcap_{i=1}^{j} \text{conv}(A_i) \neq \emptyset$. The main result of the paper is:

Theorem. Let $m = r_j(X, \mathcal{C}(X))$ and $n = r_j(Y, \mathcal{C}(Y))$. Then the following inequality holds: $\max\{m, n\} \leq r_j(X \times Y, \mathcal{C}(X) \oplus \mathcal{C}(Y)) \leq \max\{(j - 1)(\max\{m, n\} + 1) - 1, (j - 1)(\min\{m, n\} + 1) + \max\{m, n\}\}$.

Sharpness of the two bounds in the theorem is considered, and it is found that both are sharp in many situations.

TVERBERG-TYPE THEOREMS IN CONVEX PRODUCT STRUCTURES

Gerald Thompson
Augusta College

William R. Hare
Clemson University

1. Introduction

Radon's theorem [3] states that each set of $d + 2$ points in R^d can be partitioned into two nonempty subsets whose convex hulls have a point in common. Tverberg [4] found a generalization: Each set of $(j - 1)(d + 1) + 1$ points in R^d can be partitioned into j nonempty subsets whose convex hulls have a point in common. Levi [2] introduced the notion of a <u>convex structure</u> as an ordered pair $(X, C(X))$, where X is a set and $C(X) \subset 2^X$ satisfies (i) X, $\emptyset$ are elements of $C(X)$ and (ii) if $\mathcal{J} \subseteq C(X)$, then $\cap \mathcal{J} \in C(X)$. For $A \subseteq X$, conv A is the intersection of the subfamily of $C(X)$ consisting of those members which contain A. Eckhoff [1] extended the notion of Levi to that of a <u>product structure</u>: if $(X, C(X))$ and $(Y, C(Y))$ are convex structures, their structure is $(X \times Y, C(X) \oplus C(Y))$ where $C(X) \oplus C(Y) = \{A \times B \mid A \in C(Y)\}$. In this setting Eckhoff studied analogs of Radon's theorem. In the present paper a corresponding abstraction of Tverberg's theorem will be studied.

2. The j-order Radon Index

The j-order Radon index $r_j(X, C(X))$ is the smallest nonnegative integer r (if such an integer exists) such that, if $A \subseteq X$ has at least $r + 2$ elements, then A can be partitioned as $A = \bigcup_{i=1}^{j} A_i$ with $\bigcap_{i=1}^{j} \text{conv } A_i \neq \emptyset$.

3. Examples

(a) $r_j(X, \{\emptyset, X\}) = j - 2$

(b) For any set X, let $C_k(X)$ consist of all subsets of X having at most k elements, together with X itself.

Then $r_j(X, C_k(X)) = (j - 1)(k + 1) - 1$.

(c) If X is a finite set with n elements, then

$r_j(X, 2^X) = n - 1$.

(d) If C^d denotes the usual convex sets in R^d, then

Tverberg's theorem says that $r_j(R^d, C^d) = (j - 1)(d + 1) - 1$.

4. <u>Bounds for the j-order Radon Index in a Product Structure</u>

In this section the main result is proved and its relationship to Eckhoff's theorem is noted.

<u>Theorem A</u>. Let $m = r_j(X, C(X))$ and $n = r_j(Y, C(Y))$, then

$$\max\{m, n\} \leq r_j(X \times Y, \; C(X) \oplus C(Y))$$
$$\leq \max\{(j - 1)(\max\{m, n\} + 1) - 1, \; (j - 1)(\min\{m, n\} + 1)$$
$$+ \max\{m, n\}\}.$$

<u>Proof</u>. It may be assumed without loss of generality that $\max\{m, n\} = m$, thus it must be shown that

$$m \leq r_j(X \times Y, \; C(X) \oplus C(Y)) \leq \max\{(j - 1)(m + 1) - 1,$$
$$(j - 1)(n + 1) + m\}.$$

In order to prove the first inequality it suffices to show the existence of $m + 1$ points in $X \times Y$ which are not j-order Radon-decomposable. Let $A \subset X$ be such that $\mathrm{card}\, A = m + 1$ and A is not j-order-Radon-decomposable. Let $a \in Y$; it is asserted that $A \times \{a\}$ is not j-order-Radon-decomposable. If $A \times \{a\}$ were j-order-Radon-decomposable, then there would exist sets $A_1, A_2, \ldots, A_j$, in X such that

$$A \times \{a\} = A_1 \times \{a\} \cup A_2 \times \{a\} \cup \ldots \cup A_j \times \{a\}$$

with $A_i \times \{a\} \cap A_k \times \{a\} = \emptyset$ if $i \neq k$, and $\overset{j}{\underset{i=1}{\cap}} \mathrm{conv}(A_i \times \{a\}) \neq \emptyset$. Thus $A_i \cap A_k = \emptyset$ if $i \neq k$, and

$$\bigcap_{i=1}^{j} \text{conv}(A_i \times \{a\}) = \bigcap_{i=1}^{j} \text{conv } A_i \times \text{conv}\{a\} \neq \emptyset.$$

Hence, $\bigcap_{i=1}^{j} \text{conv } A_i \neq \emptyset$; but this contradicts the fact that A is not j-order-Radon-decomposable.

In order to prove the second inequality let $k =$ $\max\{(j - 1)(m + 1) - 1, (j - 1)(n + 1) + m\}$ and let $E \subset X \times Y$ with card $E = k + 2$. Now card $P_x E \geq m + 2$ or for some $b \in P_x E$ it must be the case that $\text{card}(P_x^{-1}\{b\} \cap E) \geq j$, since card $E \geq$ $(j - 1)(m + 1) + 1$.

In either case there exists a subset B of E with card $B \leq$ $m + 2$ which can be partitioned as $B_1 \cup B_2 \cup \ldots \cup B_j$ with $B_i \cap B_\ell = \emptyset$, if $i \neq \ell$, and $\bigcap_{i=1}^{j} \text{conv } P_x B_i \neq \emptyset$. Let $x \in B_1$ and $A = (E - B) \cup \{x\}$. Since card $E \geq (j - 1)(n + 1) + m + 2$ it follows that card $A \geq (j - 1)(n + 1) + 1$. Thus A can be partitioned as $A_1 \cup A_2 \cup \ldots \cup A_j$ with $A_i \cap A_\ell = \emptyset$, if $i \neq \ell$, and $\bigcap_{i=1}^{j} \text{conv } P_y A_i \neq \emptyset$. The labeling may be chosen in such a way that $x \in A_1$. Let $F_i = A_i \cup B_i$ for $i = 1, 2, \ldots, j$. It is clear that $F_i \cap F_\ell = \emptyset$ for $i \neq \ell$, and $\bigcup_{i=1}^{j} F_i = E$. It is asserted that $\bigcap_{i=1}^{j} \text{conv } F_i \neq \emptyset$; for

$$\bigcap_{i=1}^{j} \text{conv } F_i = \bigcap_{i=1}^{j} (\text{conv } P_x F_i \times \text{conv } P_y F_i)$$
$$\supset \bigcap_{i=1}^{j} \text{conv } P_x F_i \times \bigcap_{i=1}^{j} \text{conv } P_y F_i$$
$$\supset \bigcap_{i=1}^{j} \text{conv } P_x B_i \times \bigcap_{i=1}^{j} \text{conv } P_y A_i$$
$$\neq \emptyset.$$

This completes the proof of the theorem.

It should be noted that

$$\max\{(j - 1)(m + 1) - 1, (j - 1)(n + 1) + m\} = (j - 1)(m + 1) - 1$$

if and only if $(j - 2)m \geq (j - 1)n + 1$.

Corollary 1. If $m = r_2(X, C(X))$ and $n = r_2(Y, C(Y))$, then $\max\{m, n\} \leq r_2(X \times Y, C(X) \oplus C(Y)) \leq m + n + 1$.

Corollary 2. If $j = 2$ and either m or n is 0, then $r_2(X \times Y, C(X) \oplus C(Y)) = \max\{m, n\} = m + n$.

These are two of the basic theorems proved by Eckhoff in [1].

5. Sharpness of the Bounds

In this section it is shown that, for certain values of j, m and n, the bounds in Theorem A are best possible.

Example (a). If m, n and j are nonnegative integers with $j \leq \min\{m, n\} + 2$, and if both $m + 1$ and $n + 1$ are divisible by $j - 1$, then there exist convex structures $(X, C(X))$ $(Y, C(Y))$ having j-order Radin indices m and n, respectively, such that

$$r_j(X \times Y, C(X) \oplus C(Y)) = \max\{m, n\}.$$

They are defined as follows:

Let X and Y = set of positive integers and

$$C(X) = C_k(X) \quad \text{where} \quad k = \frac{m + 1}{j - 1} - 1$$

and

$$C(Y) = C_\ell(Y) \quad \text{with} \quad \ell = \frac{n + 1}{j - 1} - 1.$$

The proof of the assertion is quite lengthy, though elementary, and is omitted.

Example (b). Let

$$X = \{1, 2, \ldots, m+1\}, \quad Y = \{1, 2, \ldots, n+1\},$$

$C(X) = 2^X$ and $C(Y) = 2^Y$. Then $r_j(X, C(X)) = m$, $r_j(Y, C(Y)) = n$ and $r_j(X \times Y, C(X) \oplus C(Y)) = \min\{(j - 1)(\max\{m, n\} + 1) - 1$, $(m + 1)(n + 1) - 1\}$. Again the proof is omitted. Note that for $3 \leq j \leq \min\{m, n\} + 1 = n + 1$, $m = \max\{m, n\}$ and $(j - 2)m \geq (j - 1)n + 1$, the upper bound in Theorem A is achieved.

REFERENCES

[1] Eckhoff, J. "Der Satz von Radon in konvexen Produktstrukturen I, II," <u>Monat. für Math</u>. 72 (1968), 303-314; 73 (1969), 7-30.

[2] Levi, F.W. "On Helly's theorem and the axioms of convexity," J. Indian Math. Soc. (N.S.) Part A 15 (1951), 65-76.

[3] Radon, J. "Mengen konvexer Korper, die einem gemeinsamen Punkt enhalten," Math. Ann. 83 (1921), 113-115.

[4] Tverberg, H. "A generalization of Radon's theorem," J. Lon. Math. Soc. 41 (1966), 123-128.

INTERSECTING FAMILIES OF CONVEX COVER ORDER TWO

Marilyn Breen
University of Oklahoma

ABSTRACT

Little work has been done on the problem of characterizing intersections of certain maximal 3-convex subsets of a set, and in general,
there seems to be no 'nice' characterization for such an intersection.
In this paper, the question is examined from a different point of view.

Throughout, let $S = cl(int\ S)$ be a subset of the plane, Q the
set of points of local nonconvexity of S, Q having cardinality n,
with p a point in (bdry $S \cap$ ker S) $\sim Q$. For H a line supporting
conv S at p, let R_0, R_{n+1} be distinct closed rays at p having
union H. Finally, consider the collection of rays consisting of R_1,
R_{n+1}, together with all rays of the form $R(p,q)$ emanating from p
through q for some q in Q. We order these rays in a clockwise
direction. Consecutive rays determine a closed subset T of S, and
we call the convex set $cl(int\ T)$ a <u>wedge</u> of S.

If S is m-convex, the wedges of S play an important role in
obtaining a decomposition theorem for S, for the collection of all
wedges of S has convex cover order $m - 1$. That is, the wedges of
S may be partitioned into $m - 1$ sets $\mathcal{C}_i$ so that $conv(\cup \{V : V$ in
$\mathcal{C}_i\}) \subseteq S$ for $1 \leq i \leq m - 1$. Hence S is 3-convex if and only if
its set of wedges has convex cover order 2.

This in turn motivates the problem considered here. We will
examine those maximal families of wedges of S having convex cover
order 2, and characterize those wedges which belong to each family.

<u>Theorem</u>. If $\mathcal{M}$ denotes the collection of all maximal families
of wedges of S having convex cover order 2, then the members of $\cap \mathcal{M}$
are characterized in the following manner: For W a wedge, let V_1, V_3
denote the immediate predecessor and immediate successor of W in our
clockwise ordering (if they exist). Then W is in $\cap \mathcal{M}$ if and only

if conv($V_1 \cup V_3$) $\subseteq$ S and for every wedge $A \neq V_1, V_3$, conv($W \cup A$) $\subseteq$ S.

INTERSECTING FAMILIES OF CONVEX COVER ORDER TWO

Marilyn Breen
University of Oklahoma

Summary

Let $S = cl(int\ S)$ be a subset of the plane, Q the set of points of local nonconvexity of S, Q finite, with p a point in (bdry $S \cap ker\ S$) $\sim Q$. If $\mathcal{M}$ denotes the collection of all maximal families of wedges of S having convex cover order 2, then the members of $\cap\, \mathcal{M}$ are characterized in the following manner: For W a wedge, let V_1, V_3 denote the immediate predecessor and immediate successor of W in our clockwise ordering, if they exist. (Otherwise, either $V_1 = \emptyset$ or $V_3 = \emptyset$.) Then W is in $\cap\, \mathcal{M}$ if and only if $conv(V_1 \cup V_3) \subseteq S$ and for every wedge $A \neq V_1, V_3$, $conv(W \cup A) \subseteq S$.

1. Introduction

Let S be a subset of the plane. The set S is said to be m convex, $m \geq 2$, if and only if for every m-member subset of S, at least one of the $\binom{m}{2}$ line segments determined by these points lies in S. A point x in S is called a point of local convexity of S if and only if there is some neighborhood N of x such that $S \cap N$ is convex. If S fails to be locally convex at some point q in S, then q is called a point of local nonconvexity (lnc point) of S. Throughout the paper, conv S, ker S, bdry S, int S, and cl S will be used to denote the convex hull, kernel, boundary, interior, and closure, respectively, of the set S.

Several interesting decomposition theorems have been proved for a closed, planar, 3-convex set S. Valentine [6] has shown that such a set S may be written as a union of three or fewer closed convex sets. If, in addition, S is bounded and has some point of local convexity in bdry $S \cap ker\ S$, then S is a union of two closed convex sets by a result of Stamey and Marr [3].

However, little work has been done on the problem of characterizing intersections of certain maximal 3-convex subsets of a set. Tattersall [4] has obtained conditions under which the intersection of all maximal m-convex subsets of a set S will be exactly the kernel of S, but in general, there seems to be no 'nice' characterization for such an intersection. In this paper, the question is examined from a different point of view.

Throughout the paper, S will denote a closed subset of the plane, Q the set of points of local nonconvexity of S, with p a point in (bdry $S \cap$ ker S) $\sim Q$. We assume that Q is a finite set of cardinality $n \geq 1$, and to avoid some uninteresting special cases, assume that $S = \text{cl(int } S)$.

We begin by selecting some neighborhood N of the point of local convexity p such that $N \cap S$ is convex. Let H be a supporting hyperplane to $N \cap S$ at p, and let R_0, R_{n+1} be the corresponding closed rays at p with $R_0 \cup R_{n+1} = H$. Since p is in ker S, S lies in one of the closed halfspaces determined by H. Now consider the family $\mathcal{R}$ of rays consisting of R_0, R_{n+1}, together with all rays of the form $R(p,q)$ emanating from p through q for some q in Q. Order the rays of $\mathcal{R}$ in a clockwise direction so that (for an appropriate labeling) R_0 and R_{n+1} denote the first and last rays, respectively, in our ordering. Since $S = \text{cl(int } S)$, it is easy to see that each ray R_i in $\mathcal{R} \sim \{R_0, R_{n+1}\}$ contains a unique member q_i of Q, and hence there are n of these rays. Each pair of consecutive rays R_i, R_{i+1}, $0 \leq i \leq n$, determines a closed subset T_i of S, and we call $V_i = \text{cl(int } T_i)$ a __wedge__ of S. For convenience of notation, let $q_0 = q_{n+1} = p$. Then each wedge V_i is associated with a pair of points $v_i = q_i$, $v_i' = q_{i+1}$ in $Q \cup \{p\}$, $0 \leq i \leq n$. Clearly S is the union of its wedges. Moreover, using a lemma of Valentine [5, Corollary 1], it is easy to show that each wedge of S is convex.

We make the following definitions: Let C be a family of wedges of S. We say that C has <u>convex cover order</u> m if and only if C may be partitioned into m sets $C_1, \ldots, C_m$ (called components) so that conv($\cup \{V : V$ in $C_i\}$) $\subseteq$ S for $1 \leq i \leq m - 1$. We say that the family C of wedges is a <u>maximal</u> family having convex cover order m if and only if C has convex cover order m and whenever $\mathcal{D}$ is a family of wedges of S having convex cover order m and $C \subseteq \mathcal{D}$, then $C = \mathcal{D}$.

If S is m-convex, the wedges of S play an important role in obtaining a decomposition theorem for S, for the collection of all wedges of S has convex cover order m - 1. (Breen [1], Breen and Kay [2].) Hence S is 3-convex if and only if its set of wedges has convex cover order 2. This in turn motivates the problem considered here. Instead of studying certain maximal 3-convex subsets of S, we will examine those maximal families of wedges of S having convex cover order two, and characterize those wedges which belong to each family.

<u>Theorem</u>. Let S = cl(int S) be a subset of the plane, Q the set of points of local nonconvexity of S, Q finite, with p a point in (bdry S $\cap$ ker S) $\sim$ Q. If $\mathcal{m}$ denotes the collection of all maximal families of wedges of S having convex cover order 2, then the members of $\cap \mathcal{m}$ are characterized in the following manner: For W a wedge, let V_1, V_3 denote the immediate predecessor and immediate successor of W in our clockwise ordering, if they exist. (Otherwise either $V_1 = \emptyset$ or $V_3 = \emptyset$.) Then W is in $\cap \mathcal{m}$ if and only if conv($V_1 \cup V_3$) $\subseteq$ S and for every wedge $A \neq V_1, V_3$, conv($W \cup A$) $\subseteq$ S.

<u>Proof</u>. Let W be a wedge of S and let V_1, V_3 denote the immediate predecessor and immediate successor of W, respectively, in our clockwise ordering of wedges. In case W is the first wedge of our ordering, then $V_1 = \emptyset$, and if W is the last wedge, then $V_3 = \emptyset$.

We assume that W belongs to every family in m and prove that W satisfies the two conditions given above. We verify the second condition first. Certainly for every pair A,B of wedges of S, there is a collection in m containing A and B. Hence some member C of m contains V_1 and V_3. Since W is adjacent to both V_1 and V_3, it is clear that conv(W ∪ V_i) ⊈ S for i = 1,3. Hence W and V_i cannot belong to the same component in any partition for C, so V_1, V_3 must be in the same component, and conv(V_1 ∪ V_3) ⊆ S.

To check that the first condition is satisfied, we let A be any wedge distinct from V_1, W, V_3, and prove that conv(W ∪ A) ⊆ S. Without loss of generality, we assume that A follows W in our clockwise ordering. (If A precedes W, a parallel argument holds.) Let V_4 denote the immediate successor of V_3 in our ordering. Since V_3 and V_4 belong to some member of m, by repeating an earlier argument, conv(W ∪ V_4) ⊆ S. Then if A = V_4, we are finished. If A ≠ V_4, then since A and V_3 belong to some member of m, at least one of the sets conv(A ∪ V_3), conv(A ∪ W) lies in S. In case conv(A ∪ V_3) is not in S, the result is immediate. Thus assume that conv(A ∪ V_3) ⊆ S.

Recall that a,a′ denote the members of Q ∪ {p} corresponding to the wedge A. Clearly a,p,v_3 are distinct and by earlier remarks, the points cannot be collinear. We will show that both W and A lie in the closed halfspace cl(L_1) determined by the line L = L(a,v_3) and containing p, where L(a,v_3) denotes the line through a and v_3.

To see that W ⊆ cl(L_1), first notice that v_4 is necessarily in the opposite closed halfspace cl(L_2): Otherwise, since v_4 and a are in the same open halfspace determined by L(p,v_3), v_4 would lie interior to conv{p,a,v_3}. But conv{p,a,v_3} ⊆ conv(A ∪ V_3) ⊆ S, contradicting the fact that v_4 is a lnc point for S. Now clearly since v_3 = w′, each point of W lies in the closed halfspace determined by $L(p,v_3)$ and not containing a, and some point $\bar{w}$ of W lies in the corresponding open halfspace. Hence if any x in W were in the open halfspace L_2, then v_3 would be interior to $conv\{p,\bar{w},x,v_4\}$, impossible since $conv\{p,\bar{w}x,v_4\} \subseteq conv(W \cup V_4) \subseteq S$. Thus W does indeed lie in $cl(L_1)$, the desired result.

Using the fact that $conv(A \cup V_3) \subseteq S$, a similar argument may be used to show that $A \subseteq cl(L_1)$. Then since $[a,v_3] \subseteq S$, it is easy to see that $conv(A \cup W) \subseteq S$, finishing this part of the proof.

To prove that the converse is true we let W be a wedge of S satisfying the two conditions stated in the theorem, and show that W belongs to every member of m. Letting C be a collection of wedges having convex order 2 with $W \notin C$, we show that $C \cup \{W\}$ also has convex cover order 2. In case V_1 and V_3 are not both in C, the proof is easy. Hence assume that $V_1,V_3 \in C$. We assert that there is some partition for C such that V_1,V_3 both belong to the same component of the partition.

Let $\{C_1,\ldots,C_m\}$, $\{D_1,\ldots,D_k\}$ be components in a partition for C, where the sets are indexed so that C_i precedes C_{i+1} in our clockwise ordering of wedges, $1 \leq i \leq m-1$, and D_j precedes D_{j+1}, $1 \leq j \leq k-1$. If V_1,V_3 belong to the same component, there is nothing to prove, so for a particular i and j, let $V_1 = C_i$, $V_3 = D_j$. We will show that the sets $C_1 \equiv \{C_1,\ldots,C_i,D_j,\ldots,D_k\}$, $C_2 \equiv \{D_1,\ldots,D_{j-1},C_{i+1},\ldots,C_m\}$ are also components in a partition for C.

Since $C_i = V_1$, $D_j = V_3$, and $conv(V_1 \cup V_3) \subseteq S$, it is clear that $conv(\cup \{V:V \text{ in } C_1\}) \subseteq S$. Now examine the members of C_2. Since D_{j-1} precedes $D_j = V_3$ and $D_{j-1} \neq W$, D_{j-1} necessarily precedes W in our clockwise ordering. Similarly C_{i+1} follows $C_i = V_1$ and $C_{i+1} \neq W$, so C_{i+1} follows W. Also, $D_{j-1} \neq V_1,V_3$ and $C_{i+1} \neq V_1,V_3$, so by hypothesis $conv(D_{j-1} \cup W) \subseteq S$ and $conv(W \cup C_{i+1}) \subseteq S$. Hence it is easy to see that $conv(D_{j-1} \cup C_{i+1})$

$\subseteq$ S and conv($\cup$ {V:V in C_2}) $\subseteq$ S. Therefore C_1, C_2 are components in a partition for C, and since V_1, V_3 both belong to C_1, the assertion is proved.

However, then $C_1, C_2 \cup \{W\}$ are components in a partition for $C \cup \{W\}$, and $C \cup \{W\}$ has convex cover order 2. Thus every collection of convex cover order 2 may be extended to a collection of convex cover order 2 which contains W, and W must belong to every member of m, finishing the proof of the theorem.

REFERENCES

[1] Marilyn Breen, "A decomposition theorem for m-convex sets," sub-
     mitted to Israel J. Math.

[2] Marilyn Breen and David C. Kay, "General decomposition theorems
     for m-convex sets in the plane," submitted to Israel J. Math.

[3] W.L. Stamey and J.M. Marr, "Unions of two convex sets," Canad J.
     Math. 15 (1963), 152-156.

[4] J.J. Tattersall, "On the intersection of maximal m-convex sub-
     sets," Israel J. Math. 16 (1963), 300-305.

[5] F.A. Valentine, "Local convexity and L_n sets," Proc. Amer. Math.
     Soc. 16 (1965), 1305-1310.

[6] F.A. Valentine, "A three point convexity property," Pacific J.
     Math. 7 (1957), 1227-1235.

A HELLY-TYPE THEOREM FOR WIDTHS

G.T. Sallee

University of California, Davis

§1. Introduction

All of our work takes place in E^d, d-dimensional Euclidean space with unit sphere S^{d-1} and is concerned solely with closed convex sets. In this setting, we wish to prove a metric analogue to the well-known

Helly's Theorem. Let $K_1, \ldots, K_n$ be a collection of convex sets in E^d with the property that the intersection of every d + 1 of them is non-empty. Then the intersection of all of them is non-empty.

Recall that the width $w(K,u)$ of a closed convex set K in direction u is the distance between the pair of parallel hyperplanes normal to u which support K; $w(K,u)$ may be infinite. The width of K, $w(K)$ equals $\min\{w(K,u): u \in S^{d-1}\}$. We arbitrarily set $w(\emptyset) = -1$ where $\emptyset$ denotes the empty set. For additional information on these matters, consult [2].

Theorem 1. Let $K_1, \ldots, K_n$ be a collection of closed convex sets in E^d with the property that the width of the intersection of every d + 1 of them is at least $\alpha \geq 0$. Then $w(K_1 \cap \ldots \cap K_n) = \beta \geq \alpha$. Moreover, there exists some d + 1 of the K_i -- say $K_1, \ldots, K_{d+1}$ -- such that $w(K_1 \cap \ldots \cap K_{d+1}) = \beta$.

Note that Helly's Theorem is obtained as a special case of the above result by setting $\alpha = 0$. The result is also closely related to Vincensini's Theorem and some results of Klee. For an excellent survey of the related literature see [1].

The idea of the proof is to establish the result for polyhedra and then proceed to the general case by approximation. A preliminary lemma is needed to characterize when the minimum width occurs for polyhedra.

§2. Minimum Widths for Polyhedra

A <u>polyhedron</u> in E^d is simply the intersection of a finite collection of halfspaces. A bounded polyhedron is termed a <u>polytope</u> and is also the convex hull of a finite set of points. Let K be a closed convex set and $u \in S^{d-1}$. Then $H(K,u)$ denotes the hyperplane supporting K with outer normal u and $F(K,u) = H(K,u) \cap K$ is called a <u>face</u> of K. Let $\dim F(K,u)$ denote the dimension of the face. A polyhedron in E^d can have proper faces of dimensions from 0 to $d - 1$. For detailed information on these notions, consult [3].

Lemma 1. <u>Let</u> P <u>be a polyhedron in</u> E^d <u>of width</u> w. <u>Then for any direction</u> u <u>for which</u> $w(P,u) = w < \infty$, <u>it follows that</u> $\dim F(P,u) + \dim F(P,-u) \geq d - 1$.

Proof. Suppose the assertion is false and let u be a direction such that $w(P,u) = w$, but $\dim F(P,u) + \dim F(P,-u) \leq d - 2$. Let $H_1 = H(P,u)$, $H_2 = H(P,-u)$, $F_1 = F(P,u)$ and $F_2 = F(P,-u)$. Notice that there is at least a 2-dimensional subspace S of directions, containing u, which is orthogonal to both F_1 and F_2. Then H_1 and H_2 can rotate slightly both ways in this subspace so that they remain parallel and still support F_1 and F_2, respectively. (This is somewhat easier to see if P is projected orthogonally onto S; there, F_1 and F_2 project into vertices.) For one of the directions in which H_1 and H_2 can rotate, they come closer together but still support P -- contrary to the assumption that w was the minimum width of P.

The conclusion follows.

Lemma 2. <u>Let</u> $P_1,\ldots,P_n$ <u>be a finite collection of polyhedra in</u> E^d <u>with the property that the width of the intersection of each</u> $d + 1$ <u>of them is at least</u> $\alpha \geq 0$. <u>Then</u> $w(P_1 \cap \ldots \cap P_n) = \beta \geq \alpha$. <u>Moreover, there exist some</u> $d + 1$ <u>of the</u> P_i -- <u>say</u> $P_1,\ldots,P_{d+1}$ -- <u>such that</u> $w(P_1 \cap \ldots \cap P_{d+1}) = \beta$.

Proof. Without loss of generality, we may assume that each of the P_i is a halfspace. Let $Q = P_1 \cap \cdots \cap P_n$. By Helly's Theorem, Q is not empty since we assumed that the intersection of each $d + 1$ of them is not empty. Choose a direction u so that $w(Q,u) = w(Q) = \beta$.

Since each $d - 1$ dimensional face of Q clearly lies on the bounding hyperplane of one of the P_i, each k-face of Q lies on the intersection of some $d - k$ of the P_i. Thus, if $i_1 = \dim F(Q,u)$ and $i_2 = \dim F(Q,-u)$, we may find a collection of $m = (d - i_1) + (d - i_2)$ of the P_i such that the faces $F(Q,u)$ and $F(Q,-u)$ lie on the boundary of their intersection, R. Hence, $H(Q,u)$ and $H(Q,-u)$ support R, so $w(R) \leq \beta$. But since $Q \subseteq R$, $w(R) \geq w(Q) = \beta$, and so $w(R) = \beta$. Moreover, since $i_1 + i_2 \geq d - 1$, by Lemma 1, it follows that $m \leq d + 1$.

Hence, there exist $d + 1$ of the P_i whose intersection has width β. It follows that $\beta \geq \alpha$ and the proof is complete.

§3. Proof of Theorem 1.

Once again we appeal to Helly's Theorem to deduce that $L = K_1 \cap \cdots \cap K_n$ is not empty. It remains to show that $w(L) \geq \alpha$. If $\alpha = 0$, this, too, follows from Helly's Theorem, so we will assume $\alpha > 0$. Our strategy is to approximate the K_i by polyhedra and then use Lemma 2. We will need the observation that for any closed convex set $K \subseteq E^d$ with $w(K) < \infty$, there exists a polytope $Y \subseteq K$ such that $w(Y) \leq w(K) < w(Y) + \epsilon$ for any $\epsilon > Q$.

Now let Ω denote the set $\{1,2,\ldots,n\}$. If $\Lambda = \{i(1),\ldots, i(r)\}$ is any non-empty subset of Ω, and $X_1,\ldots,X_n$ is any set indexed by Ω, let $X(\Lambda) = X_{i(1)} \cap \cdots \cap X_{i(r)}$. Then it follows from the observation above that for any non-empty $\Lambda \subseteq \Omega$ and for any $\epsilon > 0$, there exists a polytope $Y(\Lambda,\epsilon) \subseteq K(\Lambda)$ such that $w(K(\Lambda)) < w(Y(\Lambda,\epsilon)) + \epsilon$. For each i, $1 \leq i \leq n$, let Q_i^ϵ denote the convex hull of all those $Y(\Lambda,\epsilon)$ such that $i \in \Lambda$. Clearly, for all $\Lambda \subseteq \Omega$, $Y(\Lambda, \varepsilon) \subseteq Q^\varepsilon(\Lambda) \subseteq K(\Lambda)$ and thus $w(Q^\varepsilon(\Lambda)) + \varepsilon > w(K(\Lambda)) \geq w(Q^\varepsilon(\Lambda))$.

By hypothesis, if Λ contains $d + 1$ elements, $w(Q^\varepsilon(\Lambda)) > w(K(\Lambda)) - \varepsilon \geq \alpha - \varepsilon$. Hence, by Lemma 2, $w(Q^\varepsilon(\Omega)) > \alpha - \varepsilon$. But since $L = K(\Omega)$, we have $w(L) > \alpha - \varepsilon$. Since this assertion is true for all $\varepsilon > 0$, the first conclusion follows.

For the remaining part of the proof, we know that by Lemma 2, there exists a set Λ^ε having $d + 1$ elements such that $w(Q^\varepsilon(\Omega)) = w(Q^\varepsilon(\Lambda^\varepsilon))$. Hence, $w(K(\Lambda^\varepsilon)) - \varepsilon < w(Q^\varepsilon(\Lambda^\varepsilon)) \leq w(L) \leq w(K(\Lambda^\varepsilon))$. This is true for all $\varepsilon > 0$, but possibly with a different Λ^ε. However, as $\varepsilon \to 0$, there exists some set, Λ°, which occurs infinitely often for arbitrarily small ε. Using Λ° in place of Λ^ε above and passing to the limit, we see that $w(K(\Lambda^\circ)) = w(L)$ and the proof is complete.

§4. <u>A New Proof of Helly's Theorem</u>.

The same idea as was used in the previous sections (of approximating the convex sets K_i by polyhedra $P_i \subseteq K_i$) may also be used to prove Helly's Theorem. In this case we show that $\cap P_i \neq \emptyset$ from which it clearly follows that $\cap K_i \neq \emptyset$. In fact, the proof below could be appended to the beginning of the previous one to eliminate any need to cite Helly's Theorem there, but doing so seems to obscure its structure.

So, in the notation of the previous section, for each $\Lambda \subseteq \Omega$ with exactly $d + 1$ elements, let $P(\Lambda) \subseteq K(\Lambda)$ be a non-empty polytope and let P_i denote the convex hull of all $P(\Lambda)$ with $i \in \Lambda$. It is clear that $P_i \subseteq K_i$ and by construction, the intersection of each $d + 1$ of the P_i is non-empty.

Replace each P_i by the (finite) set of halfspaces determined by its facets. Let $H_1, \ldots, H_s$ denote the set of all these halfspaces determined by the P_i. Clearly, the intersection of each $d + 1$ of the H_j is non-empty and we wish to show that the intersection of all the H_j is non-empty.

Suppose the contrary. Then for some $r \geq d$, $H_1 \cap \ldots \cap H_r = W \neq \emptyset$, but $W \cap H_{r+1} = \emptyset$. Since W is, by definition, a polyhedron, the minimum distance between W and H_{r+1} is attained for an entire e-dimensional face of W. But this e-dimensional face is determined by the intersection of some $d - e$ of the H_i, say $H_1, \ldots, H_{d-e}$ [3, p. 35]. Then, in particular $H_1 \cap \ldots \cap H_d \cap H_{r+1} = \emptyset$, contrary to hypothesis. So the intersection of all of the H_j is non-empty and Helly's Theorem is proved.

§5. Remarks

A somewhat weaker version of Theorem 1 may be extended to infinite collections of convex sets as follows.

Theorem 2. Let $\{K_\tau : \tau \in T\}$ be a collection of compact, convex sets which all lie inside some bounded set $M \subseteq E^d$. If the width of the intersection of each $d + 1$ of the K_τ is at least $\alpha \geq 0$, then $w(\cap\{K_\tau : \tau \in T\}) \geq \alpha$.

The proof of the above is easily obtained by using Theorem 1 and approximating. Note, however, that we can no longer assert that $w(\cap\{K_\tau : \tau \in T\})$ is attained as the width of the intersection of some $d + 1$ of the K_τ. For an easy example of this, let $\{u_n\}$ be a countably dense set of directions on S^1, the unit circle, such that $u_i \neq -u_j$ for any i, j. Now let K_n be the subset of the unit disc which lies in the halfspace $\{x : \langle x, u_n \rangle \leq 1/2\}$, where $\langle , \rangle$ denotes the usual inner product. It is clear that $\cap K_i$ is a disc of radius $1/2$ and thus $w(\cap K_i) = 1$, but $w(K_i \cap K_j \cap K_k) > 1$ for all i, j, k. In fact, for this example, no finite intersection has width 1.

REFERENCES

[1] L. Danzer, B. Grünbaum and V. Klee, Helly's Theorem and its
     relative, <u>Proc</u>. <u>Symp</u>. <u>Pure</u> <u>Math</u>., 7 (Convexity), 101-180
     (1963).

[2] H.G. Eggleston, <u>Convexity</u>, Cambridge, 1958.

[3] B. Grünbaum, <u>Convex</u> <u>Polytopes</u>, Wiley and Sons, 1967.

PROBLEMS

Edited by Richard K. Guy

1. (Kenneth B. Stolarsky) M.S. Klamkin (unpublished) has proved my conjecture [2] that if $a, A; b; c, C$ are the lengths of the pairs of opposite sides of a tetrahedron T with base abc, then

$$\frac{BC}{bc} + \frac{CA}{ca} + \frac{AB}{ab} \geq 1.$$

Without loss of generality it may be assumed that the vertex of T opposite the base abc is in the plane of abc; thus the problem is planar. Can one write down a "one line proof" by rearranging the equation $D = 0$ where D is the Cayley-Menger determinant for the vertices of T? The answer is "yes" if $a = b = c = 1$, for then $D = 0$ is the same as

$$BC + CA + AB = 1 + \frac{1}{2}\{(B - C)^2[(B + C)^2 - 1]$$
$$+ (C - A)^2[(C + A)^2 - 1] + (A - B)^2[(A + B)^2 - 1]\}.$$

$$D = \begin{vmatrix} 0 & 1 & 1 & 1 & 1 \\ 1 & 0 & A^2 & B^2 & C^2 \\ 1 & A^2 & 0 & c^2 & b^2 \\ 1 & B^2 & c^2 & 0 & a^2 \\ 1 & C^2 & b^2 & a^2 & 0 \end{vmatrix}$$

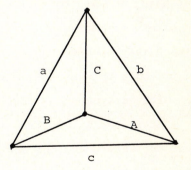

1. Leonard M. Blumenthal, Theory and Applications of Distance Geometry, Clarendon Press, Oxford, 1953, pp.97-99.

2. Kenneth B. Stolarsky, Cubic triangle inequalities, Amer. Math. Monthly 78 (1971), 879-881.

2. (Andrew Sobczyk) Geometrical realizations of colorings of simplicial complexes. The dual Ramsey number $R(s,t;r)$, for $s < r$, $t < r$ is the least n such that the K_r's of a K_n (combinatorial $(n-1)$-simplex) can be colored red and blue in such a way that no K_s of K_n is incident with only blue K_r's and no K_t of K_n is incident with only red K_r's. For example it may be easily seen that $R(1,1;2) = 4$, $R = (1,1;3) = 4$ and $R(2,2;3) = 5$. Such numbers may be generalized to any number of colors. The combinatorial situation expressed by $R(1,1;2) = 4$ may be Euclideanly realized by assigning two edge lengths a,c, e.g. in E^3 by $c = 1.1$ for both diagonals of a "square" and $a = 2$ for the sides, but evidently a geometrical constraint prevents Euclidean realization with sides $a = 1.1$ and both diagonals $c = 2$. What is the state of affairs for Euclidean (and metric) realizations of two-distance K_n's (with the requirement that each vertex have issuing from it an edge of each length) when n is greater than 4? There are similar questions for the situations $R(1,1;3) = 4$ and $R(2,2;3) = 5$, e.g., if acute and obtuse triangles (see below) are distinguished. As an example for $R(2,2;3) = 5$, may lengths be assigned for the edges of K_n for each $n \geq 5$, so that each edge is an edge of both an acute and an obtuse triangle?

The Ramsey number $R(4,4;3)$ is known to be between 12 and 18. Call a triangle obtuse if it contains an angle $\geq 90°$, otherwise acute. With assignment of possibly more than two lengths to the edges, what is the largest n such that each sub-K_4 of K_n has at least one obtuse and one acute triangle?

For K_5 with $AB = BC = CD = DE = EA = a = 1$ and $AC = CE = EB = BD = DA = c = \sqrt{2}$, is the metric space a Euclidean 4-simplex? This space has a right triangle containing each vertex with the right angle at the vertex. The n-simplex whose vertices are the origin and the unit points on the n rectangular coordinate axes in R^n and $\binom{n}{2}$ right triangles at the origin, but all the other triangles are acute.   Generally, what distributions among the vertices of the right and obtuse angles are possible for metric and Euclidean simplexes?   [Answer by B. Grünbaum for K_5:   yes; with a = 1,   the metric space is Euclidean for c between $(\sqrt{5} \pm 1)/2$.]

For K_7, the seven triangles (i, i + 1, i + 3), mod 7 form a Steiner system.   Assign edge-lengths so that each of the seven triangles is a right triangle.   Is this metric space of 7 points Euclidean?   Metrically, by suitable assignment of lengths, the triangles of Steiner systems for larger n obviously may be made all right triangles, with equal numbers of right angles at each vertex; can this be done in such a way that the metric space is Euclidean?

3.   (R. Freese)   Given any two points p, q of a connected subset S of a metric space M,   it is clear that $\exists r \in S$ such that pr = rq.   Is connectedness enough to guarantee the existence of two points $r_1, r_2 \in S$ such that $pr_1 = r_1r_2 = r_2q$?   More generally, do there exist k points $r_i \in S$ such that $pr_1 = r_1r_2 = \ldots = r_ir_{i+1} = \ldots = r_kq$?   [Klee answers this question in the negative, but a written version of the counterexample is still in preparation.]

4.   (Victor Klee)   If we choose two points at random on the unit interval then their expected distance apart is

$$\int_0^1\int_0^1 |x - y|\,dxdy = \tfrac{1}{3}.$$

Similarly if we choose three points at random in a triangle of unit area, their convex hull has an expected area of 1/12.   I conjectured that if four points were chosen at random in a tetrahedron of unit volume, the expected volume of the convex hull would by 1/60, but a Monte Carlo evaluation gave 1/59 as the nearest reciprocal of an integer!   The exact value is unknown, and the corresponding problem is also open in higher dimensions.

Victor Klee, What is the expected volume of a simplex whose vertices are chosen at random from a given convex body? Amer. Math. Monthly 76 (1969), 286-288.

5.   (Allen Freedman, via Richard Guy) Given n points in the unit square $0 \leq x, y \leq 1$, one of which is the origin, can n non-overlapping rectangles always be found, inside the square and with sides parallel to it, each rectangle having one of the given points as its lower left hand corner, so that the total area is greater than 1/2? The original conjecture was (n + 1)/2n, which is attained when the points are (i/n,i/n), $0 \leq i \leq n - 1$. This has been proved for small n. The late H.D. Ursell believed that he had a construction for large n which prevented an area greater than c/loglog n, but attempts to reconstruct this (with points on fami-lies of rectangular hyperbolas) have failed. H.T. Croft writes that the problem is reminiscent of the following paper:

Ralph Alexander, A problem about lines and ovals, Amer. Math. Monthly 75 (1968), 482-487.

6.   (David Kay) In attempting to determine the scope of G-spaces which are locally ptolemaic (satisfy the inequality $xy \cdot zw + xz \cdot yw \geq xw \cdot yz$ in all sufficiently small neighbor-hoods) the proposer constructed the following example: Let M be an open unit disk in E^2, h the hyperbolic metric defined on M (that is, $h(x,y) = |\log\{x,y;a,b\}|$, where $\{x,y;a,b\}$ is the cross-ratio of x,y with respect to a,b and a and b are the inter-sections of line $\overleftrightarrow{xy}$ with bd M) and m the taxicab metric (if $x = (\xi_1,\xi_2)$, $y = (\eta_1,\eta_2)$ in terms of a rectangular coordinatiza-tion of E^2, then $m(x,y) = |\xi_1 - \eta_1| + |\xi_2 - \eta_2|$). It is clear that $xy = h(x,y) + m(x,y)$ defines a metric for M which will make the chords of M the geodesics.

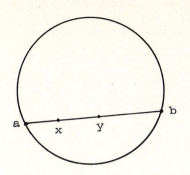
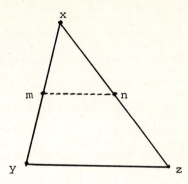

In fact M is an example of a straight G-space, and it is easily

shown that M is not locally ptolemaic anywhere.

Problem: Determine whether M has non-positive curvature (in

each sufficiently small neighborhood if x,y,z are any three points

with m and n the midpoints of segments $\overline{xy}$ and $\overline{xz}$ then

$2mn \leq yz$). A few calculations tediously carried out tend to answer

the question in the affirmative. If this were indeed the case, in

addition to showing that non-positive curvature is not strong enough

to imply the local ptolemaic property, M would provide a rare in-

stance of a non-Riemannian G-space with non-positive curvature. As

far as the proposer knows, no example of such a G-space has ever been

explicitly constructed.

Note: A G-space is a finitely compact (therefore complete) me-

tric space which is internally convex, externally convex in all

sufficiently small neighborhoods, and satisfies the axiom of unique

prolongability: if xyz_1, xyz_2 and $yz_1 = yz_2$ then $z_1 = z_2$.

7. (J.J. Seidel) Does there exist a graph on 99 points in

which each edge belongs to a unique triangle and each non-edge (pair

of non-adjacent points) is the diagonal of a unique quadrilateral?

The valence would be 14, with 7 triangles at each vertex, 231

triangles and 693 edges in all. [J.H. Conway and the editor have shown that the automorphism group of such a graph contains no elements of order 5,7,9 or 11. Since 99 is not a prime power, the editor conjectures non-existence.] There is such a graph on 3^2 points (join 2 vertices just if they are in the same 'row' or 'column') and also one on 3^5 points. There are two possible larger numbers for which the problem is also unsettled.

E.R. Berlekamp, J.H. van Lint and J.J. Seidel, A strongly regular graph derived from the perfect ternary Golay code, in A Survey of Combinatorial Theory, North-Holland, 1973, 25-30.

8.   (Tom Sallee) The width of a convex set is the minimum distance between a pair of parallel support planes. What is the simplex of maximum width which can be inscribed in the unit ball in E^d? It seems certain that the answer is the regular simplex, although this has not been proved even for $d = 3$. [H.T. Croft says that the techniques used in proving the relations between R and D, r and d (H.G. Eggleston, Convexity, Cambridge Univ. Press, 1953, 111-115) may be relevant. Ralph Alexander sends a proof for $d = 3$, based on the lemma: Let p_i, q_j, $1 \leq i, j \leq n$ be points in a Euclidean space, then

$$\sum_{i<j} |p_i - p_j|^2 + \sum_{i<j} |q_i - q_j|^2 - \sum_{i,j} |p_i - q_j|^2 = -(ns)^2,$$

where s is the distance between the centroids of the p_i and of the q_j. Both the lemma and the proof generalize to all dimensions, and will appear in a forthcoming paper.]

9.   (Tom Sallee) A converse to the Brunn-Minkowski theorem. Let X be a compact set in E^d, and let $L(u)$ be the line in direction u through the origin. If x represents the directed distance on $L(u)$ from the origin, let $H(x,u)$ be the hyperplane normal to u and at a distance x, and let $f_u(x) = |V(H \cap x|^{1/(d-1)}$ where V is the $(d - 1)$-dimensional volume. If $f_u(x)$ is a concave function for every direction u, what other conditions are needed on X to guarantee that X is convex? For $d = 2$ it is suggested that this is close to Hammer's X-ray problem. H.T. Croft writes that even

more relevant is the problem of Bonnice, quoting the results of

Busemann and Ewald.

William Bonnice, Problem 2, Proc. Colloq. Convexity, Copenhagen 1965,
308-309.

P.C. Hammer, Proc. A.M.S. Symp. Pure Math. 7 (Convexity), 1963, 498.

10. (G.D. Chakerian) Any plane convex set of unit area admits
a circumscribed rectangle of area at most 2. This is best possible
since every rectangle containing a triangle of unit area has area at
least 2. It can be shown that any plane convex set of unit area
admits a circumscribed quadrilateral of area at most $\sqrt{2}$. Is this
result best possible? Is there a plane convex set of unit area all
of whose circumscribed quadrilaterals have area at least $\sqrt{2}$?

11. (L.M. Kelly) _Metric detrminacy_. There is no rank order of
the six lengths of the edges of a tetrahedron which cannot be realized in E^3; in fact, if equalities are excluded, any rank order of
distances between 4 points can be realized in E^2, but 4 points
can't be found in E^2 all of whose six distances are equal.
Characterize the cases that can't be done. More generally we can
find d + 2 points in E^d, but all the triangles can't be long-
legged isosceles. Can we go to d + 3 points, if we exclude
equalities?

12. (R.B. Eggleton, A.S. Fraenkel, Richard K. Guy) Is there a
point in E^2 at a rational distance from each corner of the unit
square?

13. (Branko Grünbaum) Let $S = \{x \in X: \|x\| = 1\}$ be the unit
sphere of a normed space X and let $S^\Delta = \{x + y : y, x - y \in S\}$. Is
S an ellipsoid if it is known that S^Δ is an ellipsoid, or if $S^\Delta =$
αS for some constant α?

14. (Branko Grünbaum) Let S, X be as in 13, and let f(x, y)
be a symmetric bilinear functional and $\hat{f}(x) = f(x, x)$ the associated
quadratic functional. Their norms are $\|f\| = \sup_{x, y \in S} |f(x, y)|$ and

$\|\hat{f}\| = \sup_{x \in S} |\hat{f}(x)|$, and it is well known that $\|\hat{f}\| \leq \|f\| \leq 2\|\hat{f}\|$.

Define $\beta(X) = \sup\|f\|/\|\hat{f}\|$, where f ranges over all symmetric bi-linear functionals on X. It is well known that $\beta(X) = 1$ if X is an inner product space. Does $\beta(X) = 1$ characterize inner product spaces? More generally, what is the relation between $\beta(X)$ and the geometry of S? Do the results generalize to multilinear functionals on X?

15. (Ralph Alexander) Let C be a closed curve of length π. For each point $x \in C$, let $\delta_x = \max|x - y|$ ranges over C. We conjecture that $\min \delta_x \leq 1$ with equality just if C bounds a plane convex body of constant width 1. Results of Sallee [Geom. Dedicata, 2 (1973) 311-315] show that we need only consider curves which bound plane convex sets. If C is a square, many of the nasty properties of δ_x become apparent.

16. (James Chew) If R^n carries a multiplicative structure such that multiplication is continuous, then R^n is a topological group; i.e. the group inversion operation is also continuous. In fact it is a well known result of Ellis that if (X,T) is a topological space and X is a group with continuous multiplication, then (X,T) is a topological group provided T is a locally compact topology. It is of interest to find out what other single properties possessed by the euclidean topology imply the continuity of inversion. If X is the reals, T has $\{[a,b):a < b\}$ for a base and the group operation is ordinary addition, we have an example where inversion is not con-tinuous. Here T is paracompact but not metric. It seems natural to ask: Let (X,d) be a metric space such that X carries a multi-plicative group structure with continuous multiplication; is inver-sion continuous? I.e., is the system a topological group? [This problem has appears as Advanced Problem 5977, Amer. Math. Monthly, 81 (1974) 670, without an asterisk; presumably a solution is forthcoming.]

17. (M. Edelstein) Let S be a finite non-coplanar subset of E^3 and suppose that S is 2-colored. Does there always exist a monochromatic line (that is a straight line ℓ which contains at least two points of S and such that all points $\ell \cap S$ are of the same color)? B. Grünbaum reports that this problem, and others were solved in the following thesis:

R.W. Shannon, Certain extremal problems in arrangements of hyper-planes, Ph.D. thesis, Univ. of Washington, Seattle, 1974.

18. (J.B. Kruskal) If we have p unit vectors in d-space, no two of which are separated by more than $\pi/2$, how short can their sum be? William Kruskal conjectures that the minimum, m, is given by r groups of size $k + 1$ and $d - r$ groups of size k, where $p = kd + r$, $0 \le r \le d$. It can be shown that $p^2 \le m^2 d \le p^2 + r(d - r)$.

J.B. Kruskal and H.S. Witsenhausen, An inequality for positively correlated variables, J. Amer. Statist. Assoc., 69 (1974) 540-542.

19. (Dorothy Wolfe) Let $p_1, \ldots, p_r$ be points in a normed linear space lying on the boundary of the unit sphere and containing the origin within their convex cover. Is it true that the length of any Hamiltonian circuit of these points is at least 4? The statement is true in the Euclidean plane.

G.D. Chakerian and M.S. Klamkin, Minimal covers for closed curves, Math. Mag. 46 (1973), 55-61.

Those of us who attended the 1966 East Lansing conference will recall that the late Leo Moser circulated fifty "poorly formulated unsolved problems of combinatorial geometry". In Leo's memory and because most of the problems are still of interest, I append an edited selection; the numbers are taken from the original typescript and pre-fixed by M.

M1-2. The set of squares of sides $d_1 \ge d_2 \ge \ldots$, $\sum d_i^2 = 1$ can be placed without overlap in a single square of side $d_1 + \sqrt{1 - d_1^2} \le \sqrt{2}$. Can one obtain better estimates involving more, perhaps all, of the d_i? Extend the result to n dimensions.

M3. Can every set of rectangles of total area 1 and maximal side 1 be packed in a square of area 2?

M4. Can the 1/n by 1/(n + 1) rectangles, n = 1,2,3,..., be packed in a unit square?

M5. Can every set of rectangles of largest side 1 and total area 3 be used to cover a unit square, without rotations?

M6. The squares of sides 1/n, n = 2,3,4,..., will pack into a square of side 5/6. Will they pack into some rectangle of area $\pi^2/6 - 1$?

M7. What is the least A such that every set of squares of total area 1 can be packed into some rectangle of area A?.

M9-11. The worm problem. What is the region of smallest area which will accommodate every arc of length 1? Will the semi-ellipse $12x^2 + 16y^2 = 3$ accommodate every arc of length 1? [This has been answered negatively.] What is the largest number $f(a,b,c)$ such that every closed curve of length f can be accommodated in the triangle (if it exists) of sides a,b,c? How is it for arcs?

A.S. Besicovitch and R. Rado, A plane set of measure zero containing circumferences of every radius, J. London Math. Soc. 43 (1968) 717-719; MR 37 #5345.

J.R. Kinney, A thin set of circles, Amer. Math. Monthly, 75 (1968) 1077-1081.

J. Schaer, The broadest curve of length 1, Univ. of Calgary Math. Res. Paper #52, May 1968.

John E. Wetzel, Sectorial covers for curves of constant length, Canad. Math. Bull., 16 (1973) 367-375.

M14. Given n points on a sphere. It is conjectured that the same distance can occur at most $3n - 6$ times.

M15. What is the region of largest area which can be moved round a right angle in a corridor of width 1?

M16. If P_i, $1 \leq i \leq n$ are points on a unit sphere, then $\sum_{i \leq j} P_i P_j^2 \leq n^2$ with equality just if the centroid of the points is the centre of the sphere. Find a corresponding inequality for $\sum P_i P_j$.

M17.   If P_i, $1 \leq i \leq n$ are points with mutual distances $P_i P_j \geq 2$,   then Blichfeldt has shown that for any point   O, $\sum OP_i^2 \geq 2n - 2$ with possible equality for $n = 2,3$ and 4.   Find a corresponding sharp inequality for somewhat larger values of   n.

M19-20.   Prove that there is a function $f(n)$ which tends to infinity with   n   such that every region of area   n   can be placed on the square lattice, with rotations allowed, so as to cover $n + f(n)$ lattice points.   Also obtain good estimates for the largest such $f(n)$.

M21-23.   If all faces of a convex polyhedron have central symmetry, so has the polyhedron; can one give some corresponding result for surfaces?   At least 8 vertices of such a polyhedron are of order 3; if there are   n   vertices, at least how many must have order 3?   At least 6 faces of such a polyhedron are parallelograms; if there are   F   faces, at least how many are parallelograms?

M24.   Is there a "non-pathological" dissection of the plane, considered as a point set, into   n   congruent connected pieces?

M25.   Estimate the "size" of the largest measurable point set in a large square, which does not determine unit distance.

M26.   Given a polygon such that the smallest angle determined by three of its vertices is θ,   it can be shown that there is a fixed $c > 0$ such that the polygon can be "dissected into a square" in fewer than $c^{1/\theta}$ triangular pieces.   Improve this estimate.

M27.   What is the minimum number of pieces into which a cube can be dissected to give a cube in another orientation by translations only?

M28.   How many regular n-dimensional simplexes of edge 1 can be packed in the n-dimensional unit cube?

M32.   Dissect the surface of a sphere by   n   great circles, no 3 concurrent, so as to minimize the sum of the squares of the areas of the regions.

M35. Estimate the largest $f(n)$ such that every convex polyhedron with n vertices has an orthogonal projection on to a plane with $f(n)$ vertices on the boundary.

M37. Let $f(P) = \prod_{i=1}^{n} PP_i$ where the P_i are n points in 3-space. Give a "geometric" proof that $f(P)$ cannot have local maxima.

M41. Let $f(n)$ be the maximum number of points which can be found in the unit n-dimensional cube so that all mutual distances are at least 1. $f(n) = 2^n$ for $n = 1,2,3$. A. Meir has shown $f(4) = 17$ and several that $\log f(n) \sim \frac{1}{2}n \log n$. Evaluate $f(5)$ and sharpen the asymptotic relation.

M42. Any five or more great circles, no 3 concurrent, determine a spherical polygon of at least five sides. Sharpen or extend this result.

M49. What is the largest $f(n)$ such that every convex polyhedron with n vertices has a simple path along edges passing through $f(n)$ of the vertices? Moon and Moser have shown that $f(n) < cn^{\log 2/\log 3}$.

M50. Can every closed curve of length 2π be accommodated in a rectangle of area 4?

Vol. 457: Fractional Calculus and Its Applications. Proceedings of the International Conference Held at the University of New Haven, June 1974. Edited by B. Ross. VI, 381 pages. 1975.

Vol. 458: P. Walters, Ergodic Theory – Introductory Lectures. VI, 198 pages. 1975.

Vol. 459: Fourier Integral Operators and Partial Differential Equations. Proceedings 1974. Edited by J. Chazarain. VI, 372 pages. 1975.

Vol. 460: O. Loos, Jordan Pairs. XVI, 218 pages. 1975.

Vol. 461: Computational Mechanics. Proceedings 1974. Edited by J. T. Oden. VII, 328 pages. 1975.

Vol. 462: P. Gérardin, Construction de Séries Discrètes p-adiques. »Sur les séries discrètes non ramifiées des groupes réductifs déployés p-adiques«. III, 180 pages. 1975.

Vol. 463: H.-H. Kuo, Gaussian Measures in Banach Spaces. VI, 224 pages. 1975.

Vol. 464: C. Rockland, Hypoellipticity and Eigenvalue Asymptotics. III, 171 pages. 1975.

Vol. 465: Séminaire de Probabilités IX. Proceedings 1973/74. Edité par P. A. Meyer. IV, 589 pages. 1975.

Vol. 466: Non-Commutative Harmonic Analysis. Proceedings 1974. Edited by J. Carmona, J. Dixmier and M. Vergne. VI, 231 pages. 1975.

Vol. 467: M. R. Essén, The Cos $\pi\lambda$ Theorem. With a paper by Christer Borell. VII, 112 pages. 1975.

Vol. 468: Dynamical Systems – Warwick 1974. Proceedings 1973/74. Edited by A. K. Manning. X, 405 pages. 1975.

Vol. 469: E. Binz, Continuous Convergence on C(X). IX, 140 pages. 1975.

Vol. 470: R. Bowen, Equilibrium States and The Ergodic Theory of Anosov Diffeomorphisms. IV, 108 pages. 1975.

Vol. 471: R. S. Hamilton, Harmonic Maps of Manifolds with Boundary. IV, 168 pages. 1975.

Vol. 472: Probability-Winter School. Proceedings of the Fourth Winter School on Probability Held at Karpacz, Poland, January 1975. Edited by Z. Ciesielski, K. Urbanik, and W. A. Woyczyński. VI, 283 pages. 1975.

Vol. 473: D. Burghelea, R. Lashof, and M. Rothenberg, Groups of Automorphisms of Manifolds. (with an appendix by E. Pedersen) VII, 156 pages. 1975.

Vol. 474: Séminaire Pierre Lelong (Analyse) Année 1973/74. Edité par P. Lelong. VI, 182 pages. 1975.

Vol. 475: Répartition Modulo 1. Actes du Colloque de Marseille-Luminy, 4 au 7 Juin 1974. Edité par G. Rauzy. VI, 258 pages. 1975.

Vol. 476: Modular Functions of One Variable IV. Proceedings 1972. Edited by B. J. Birch and W. Kuyk. V, 151 pages. 1975.

Vol. 477: Optimization and Optimal Control. Proceedings 1974. Edited by R. Bulirsch, W. Oettli, and J. Stoer. VII, 294 pages. 1975.

Vol. 478: G. Schober, Univalent Functions – Selected Topics. V, 200 pages. 1975.

Vol. 479: S. D. Fisher and J. W. Jerome, Minimum Norm Extremals in Function Spaces. With Applications to Classical and Modern Analysis. VIII, 209 pages. 1975.

Vol. 480: X. M. Fernique, J. P. Conze et J. Gani, Ecole d'Eté de Probabilités de Saint-Flour IV–1974. Edité par P.-L. Hennequin. XI, 293 pages. 1975.

Vol. 481: M. de Guzmán, Differentiation of Integrals in R^n. XII, 226 pages. 1975.

Vol. 482: Fonctions de Plusieurs Variables Complexes. Séminaire François Norguet 1974–1975. IX, 366 pages. 1975.

Vol. 483: R. D. M. Accola, Riemann Surfaces, Theta Functions, and Abelian Automorphisms Groups. III, 105 pages. 1975.

Vol. 484: Differential Topology and Geometry. Proceedings 1974. Edited by G. P. Joubert, R. P. Moussu, and R. H. Roussarie. IX, 287 pages. 1975.

Vol. 485: J. Diestel, Geometry of Banach Spaces – Selected Topics. XI, 282 pages. 1975.

Vol. 486: S.-V. Stratila and D.-V. Voiculescu, Representations of AF-Algebras and of the Group U (∞). IX, 169 pages. 1975.

Vol. 487: H. M. Reimann und T. Rychener, Funktionen beschränkter mittlerer Oszillation. VI, 141 Seiten. 1975.

Vol. 488: Representations of Algebras, Ottawa 1974. Proceedings 1974. Edited by V. Dlab and P. Gabriel. XII, 378 pages. 1975.

Vol. 489: J. Bair and R. Fourneau, Etude Géométrique des Espaces Vectoriels. Une Introduction. VII, 185 pages. 1975.

Vol. 490: The Geometry of Metric and Linear Spaces. Proceedings 1974. Edited by L. M. Kelly. X, 244 pages. 1975.

This series aims to report new developments in mathematical research and teaching – quickly, informally and at a high level. The type of material considered for publication includes:

1. Preliminary drafts of original papers and monographs
2. Lectures on a new field, or presenting a new angle on a classical field
3. Seminar work-outs
4. Reports of meetings, provided they are
   a) of exceptional interest and
   b) devoted to a single topic.

Texts which are out of print but still in demand may also be considered if they fall within these categories.

The timeliness of a manuscript is more important than its form, which may be unfinished or tentative. Thus, in some instances, proofs may be merely outlined and results presented which have been or will later be published elsewhere. If possible, a subject index should be included. Publication of Lecture Notes is intended as a service to the international mathematical community, in that a commercial publisher, Springer-Verlag, can offer a wider distribution to documents which would otherwise have a restricted readership. Once published and copyrighted, they can be documented in the scientific literature.

Manuscripts

Manuscripts should comprise not less than 100 pages.
They are reproduced by a photographic process and therefore must be typed with extreme care. Symbols not on the typewriter should be inserted by hand in indelible black ink. Corrections to the typescript should be made by pasting the amended text over the old one, or by obliterating errors with white correcting fluid. Authors receive 75 free copies and are free to use the material in other publications. The typescript is reduced slightly in size during reproduction; best results will not be obtained unless the text on any one page is kept within the overall limit of 18 x 26.5 cm (7 x 10½ inches). The publishers will be pleased to supply on request special stationery with the typing area outlined.

Manuscripts in English, German or French should be sent to Prof. A. Dold, Mathematisches Institut der Universität Heidelberg, 69 Heidelberg/Germany, Tiergartenstraße, Prof. B. Eckmann, Eidgenössische Technische Hochschule, CH-8006 Zürich/Switzerland, or directly to Springer-Verlag Heidelberg.

Springer-Verlag, D-1000 Berlin 33, Heidelberger Platz 3
Springer-Verlag, D-6900 Heidelberg 1, Neuenheimer Landstraße 28–30
Springer-Verlag, 175 Fifth Avenue, New York, NY 10010/USA

ISBN 3-540-07417-1
ISBN 0-387-07417-1